World Philosophy Series

Philosophy of Science since Bacon

Readings in Ideas and Interpretations

WORLD PHILOSOPHY SERIES

Additional books in this series can be found on Nova's website under the Series tab.

Additional E-books in this series can be found on Nova's website under the E-book tab.

WORLD PHILOSOPHY SERIES

PHILOSOPHY OF SCIENCE SINCE BACON

READINGS IN IDEAS AND INTERPRETATIONS

AHMAD RAZA
EDITOR

Nova Science Publishers, Inc.
New York

For permission to use material from this book please contact us:
Telephone 631-231-7269; Fax 631-231-8175
Web Site: http://www.novapublishers.com

Library of Congress Cataloging-in-Publication Data

Raza, Ahmad, 1965-
Philosophy of science since Bacon : readings in ideas and interpretations / Ahmad Raza.
p. cm.
Includes index.
ISBN 978-1-61209-997-2 (hardcover)
1. Science--Philosophy. I. Title.
Q175.R3317 2011
501--dc22

Published by Nova Science Publishers, Inc. † New York

CONTENTS

Preface

The present volume is a collection of original articles of high quality on the Philosophy of Science by philosophers and scientists of international repute. As the title of the book suggests, it looks at the various points of view of leading practitioners, as well as philosophers on the nature and structure of our knowledge of the physical world. The present work brings forth the fundamental ideas of Bacon, Galileo, Newton, Descartes, Popper, Einstein, Thomas Kuhn and several other leading scientists and philosophers.

The collection consists of core ideas of science and how they brought about revolution in our ways of looking at knowledge and the structure of physical reality. After going through the collection, the reader can get direct experience of the workings of the great minds of science and the changing historical context of theorizing in science. It traces the evolution of human thought from Newtonian objectivity to Einsteinian relativity.

FOREWORD

The present volume is a collection of original articles of high quality on the Philosophy of Science by philosophers and scientists of international repute. As the title of the book suggests, it looks at the various points of view of leading practitioners as well as philosophers on the nature and structure of our knowledge of physical world. The present work brings forth the fundamental ideas of Bacon, Galileo, Newton, Descartes, Popper, Einestein, Thomas Kuhn and several other leading scientists and philosophers.

The collection consists of core ideas of science and how they brought about revolution in our ways of looking at knowledge and the structure of physical reality. After going through the collection, the reader can get direct experience of the workings of the great minds of science and the changing historical context of theorizing in science. It traces the evolution of human thought from Newtonian objectivity to Einsteinian relativity.

The present book can be a good resource for the academicians as well as those who want to get a deeper insight into the nature and structure of physical sciences. The book in particular is very useful for graduate students in various domains of social, natural and management sciences. It will equip them with necessary conceptual skills to analyze the issues and problems being faced currently by human societies, whose roots lie in the past theories and concepts. It will broaden their analytical and critical horizons and help them better learn and develop essential corpus of concepts, in order to achieve their research goals.

Ahmad Raza
Research Fellow, University of Management and Technology
Lahore
April 22, 2009

Chapter 1

THE STRUCTURE OF SCIENTIFIC KNOWLEDGE

Ahmad Raza

I. CLASSICAL CONCEPTION OF SCIENTIFIC KNOWLEDGE

Liberty, fraternity and equality, were the catchwords which inspired the dawn of the "Reign of Man" in 1789 and unleashed the profound creative energy which has since transformed our world to an unrecognizable extent. Man's ideas of the self and the world around were completely transformed. The "temple of reason" was resurrected with a political revolution and consequently a new form of social and philosophical life took shape, in fact a new power of mind was realized by man, the power of science. Subsequently, the natural philosophy which emerged, questioned the Medieval and Renaissance conceptual legitimacy of notions such as authority, self-image, knowledge and reality. The dissolution of old structures of perception was replaced by new method of understanding the world and its constituents. This new method of knowing was galvanized by Galileo, Kepler and Newton. The characteristic features of this method were abstraction (rational constructions), universality (generalization), observation (experimentation) and domination (control) over nature (natural world) by man.

This was the time from which, onward, civilization entered into democratic consciousness of itself and its constituents (man and culture), and autocratic notions of truth and knowledge ware abandoned forever. The logical consequence of this tremendous cosmological shift, though, appears to be simple yet contains in it profound capability to influence the traditional perceptual structures and value systems of human societies.

This new found capability to understand, observe and question the nature, structure and growth of the world, universe and it's manifestations, in a non-religious way, and then formulate rational explanations/interpretations of human existence and physical reality, constitutes the core of this classical method of knowing the world. The enthusiasm which bewitched every thinking being of 18th and 19th century Europe was the subtle understanding of the near perfect rationality of the behavior of nature, universe and the human society. This sense of profundity was the product of the power of this new method of knowing, the infallible belief in the systematic observability and universal validity of the natural world (and the laws thus formulated to describe its structure), consolidated with the conceptual tools of observation, abstraction and experimentation. The scheme, which thus emerged, perceived reality/nature as an object, an automation to be analyzed in bits and pieces, to find out it's eventual universal structure. It was operated upon, slashed, and reduced to minimum physical and mathematical terms.

The knowledge, or more appropriately scientific knowledge which then crystallized, became a kind of apparent truth, self contained and self-poised, which consistently and at times harshly separated being from becoming, man from nature, progress tram values and matter from mind. The grammar of this new methodology was thus; that the universe (and human societies included) was viewed as a big automaton, a machine, behaving in a deterministic, timeless and ordered fashion.

Change, process, growth and time-oriented activity were banned in the new found crystal palace of physics and mathematics. Thus, what, Prigogine and Stengers term as the "mechanistic world-view", triumphed over other forms of knowledge and conceptualizations of reality in the new cosmology of things. This triumph of mechanistic world-view was necessitated and supported by the rise of factory civilization. Alvin Toffler has pointedly underscored the historic coincidence between the rise of factory civilization and the triumph of mechanical weltanschauung, in his foreword to "Order out of Chaos". He says "mechanistic view coincided with the rise of factory civilization. And divine dice-shooting seems hardly enough to account for the fact that the age of the Machine enthusiastically embraced scientific theories that picture the entire universe as a machine".

Henceforth, the succeeding generations of scientists focused their energies to discover new universal "laws of nature" to be revered with awe and enforced through the selective manipulation of scientific institutions as a sacred creed, not to be challenged and questioned, to the exclusion of all other modes of scientific inquiries from this privileged status. So much so that the theoretical constructions of physics and mathematics became the criterion of scientific knowledge. This attitude of selective manipulation of scientific knowledge and its related fields of inquiries was even reflected to minute levels of study. For example the studies of certain type of phenomena and facts were suppressed and discouraged on the behest of prevalent "scientific fashions" in the institutions, and scientists (who is a kind of seeker after new facts, new phenomena, new truths), were reluctant to indulge into such studies for fear of being dubbed as backward. So it seems that the values which motivated the quest for an empirical-rational construction of the natural world were at times negated by the dominant scientific culture of the scientific institutions.

Science (and knowledge produced by its technique) is a social activity, rooted in the historical experience of mankind, inspired and influenced by the peculiar cosmological-philosophical convictions and emotional idiosyncrasies of individuals who subscribe to its methodology, and larger social context in which this activity is going on. Be it Galileo and Kepler, Newton or Einstein, Heisenberg and Boltzmann, their quest for a reasonable description of nature and reality reflects their own prejudices, choices and the prejudices and choices of their Age, in which they happen to find themselves situated and -- tormented. Their interpretations of the structure of reality are constrained by their times and traditions; in short a vision of nature underlies their conceptual frameworks.

Newtonian world-view of science also presented "a vision of nature that would be universal, deterministic and objective in as much as it contains no reference to the observer, complete in as much as it obtains a level of description that escapes the clutches of time." This vision of nature/reality, in which nature appears to be a silent and mutant version of an ancient city, wherein man is a spectator, an alienated being, surrounded by objects and tools which he has to master at any cost (even at the cost of his soul). Lonely, dehumanized and trapped in his quest for simplicity, objectivity, man experienced himself as a Kafka's character. The consequence of his double alienation, from his soul and from nature.[7] This sense of alienation and dichotomy resulted from an over confidence in the healing power of scientific knowledge and technological advancement, which was, in fact, not a pure revelation of human mind but an equally treacherous path to self-knowledge, like of which man has traversed earlier in history. The diachotomic consciousness became so entrenched that, Prigogine has to remark in a critical tone, "For ancients, nature was a source of wisdom. Medieval nature spoke of God. In modern times nature has been so silent that Kant considered that science and wisdom, science and truth, ought to be completely separated. We have been with this dichotomy for the past two centuries."

II. Tacit Component, Dissipative Structures and Anything Goes?

"We now understand that we live in a pluralistic world" and "for humans, reality is embedded in the flow of time". Similarly any attempt to describe scientific knowledge in an impersonal and objective way, which is devoid of tacit component of scientist's participation and emotional involvement in the act of knowing is barren and doomed to wither away. Laplacean ideal of scientific knowledge and the "philosophic movement guided by aspirations of scientific severity has come to threaten the position of science itself, a passion for achieving absolutely impersonal knowledge presents us with a picture of the universe in which we ourselves are absent". The varied studies in what Prigogine terms as the "science of complexity" has exposed the insufficiency of traditional epistemological assumptions of Newtonian conception of knowledge, and which is characteristically reflected in the "Laplacean delusion". Quantum mechanics, biological sciences and thermodynamics have produced new conceptual orders vis-à-vis man's understanding of diverse evolving languages, at his disposal for the description of the structure of physical reality as well as living world. "No single theoretical language can exhaust the physical contents of a system. Various possible languages and points of view about the system may be complementary." Classical world of timelessness, order and reversibility, which is described by an "objective observer", does not hold ground anymore. "The choice one makes about what he observes makes an irretrievable difference in what he finds. The observer is elevated from "observer'" to "participator". What philosophy suggested in times past, the central features of quantum mechanics tells us today with an impressive force. In some strange sense this is a participatory universe. The classical vision of nature "is undergoing a radical change toward the multiple, the temporal, and the complex" and in addition to deterministic processes, there must be an element of possibility involved in some basic processes, such as, for example, biological evolution or the evolution of human cultures. Classificatory and hierarchical theoretical language of classical world is now replaced and at times enriched by new set of concepts, describing more accurately our changing conceptualizations of reality. New concepts such as fluctuations, creativity, irreversible processes, active self-organization of matter, commitment, participation in the act of knowing, plurality of perspectives, are extensively employed by contemporary scientists and philosophers to describe the structure of scientific knowledge. It seems now that man is no more a passive subject receiving only sensory data of an all-powerful universal reality and organizing them into an "economy of thought", into descriptive tools in the service of some awesome external reality, rather his act of knowing is "dwelling in and breaking out" of his experience of reality, a kind of active participation, facilitated by these conceptual tools, an engagement in the form of a "dialogue with nature" and not it's enslavement or being enslaved, in short, he is a "macroscopic being embedded in this physical world".

The contemporary advancements in the methodologies to understand the structure of scientific knowledge are as diversely constituted as the varied temperaments of their respective designers, reflecting the multiplicity of descriptive languages and the inherent complexity of conceptual explication. This is because they are motivated by different kind of intellectual passions. Similar is the case with scientist who is engaged in the production of scientific knowledge in the first place. Ranging from Popper's conjectures and refutations (falsificationism) to Lakatos's methodology of research programs, to Feyerabend epistemological anarchism to Polanyi's personal knowledge to Prigogine's science of complexity, all of this demonstrate the theoretical diversity vis-à-vis, the "wealth of reality", which each one of them is trying to describe in his own style and grammar.

Knowledge (and the scientific knowledge, whose structure they are describing) is no more a pure activity but a human activity; which reflects in its development the power and limitations of human condition. Furthermore, it is also doubtful whether a universalistic and singular status can be accorded to scientific knowledge in the history of human thought and culture. However two aspects emerge quite clearly from these diverse methodologies; firstly, if one follows a certain type of methodological rule for understanding the nature, structure and growth of scientific knowledge, he is bound to end up with barren analysis. And secondly, man is involved in this process of acquiring new understanding, of knowing the structure of nature and reality. In fact, this endeavor is not impersonal and objective, as some would like to see it, rather it is a voyage of intimate encounters with reality, of "intellectual passion" of "unjustified (and un justifiable) anticipations" of "irreducible plurality of perspectives on the same reality" (or perhaps a different reality, a new reality), of "anything goes". Therefore, it is evident that, the sort of knowledge which the philosopher of science is trying to describe with his conceptual tools cannot be confined within a single methodological language. Simultaneously, the sort of knowledge, which is produced by "skills" called scientific, appears to be one of the models of communicating with reality and displays a structural-logical quality different from other modes of knowledge production by man. Furthermore, this mode of knowledge production may have different tactical variations within its functional fields of inquiry, motivated by strategies of differing philosophical and moral overtones of its own practitioner (scientist). Then, how can a methodologist generalize scientific knowledge as a paradigm case for all forms of human knowledge? How can he plead the case of "scientific rationality" as a model of human rationality and reduce the multiplicity of other modes of human inquiry such as arts, religious experience and myth, which are required to follow methodological constructions and rules of a rationality, whose nature, structure and sources are entirely different from these human activities? Therefore, one has to consider, if one has to have a satisfactory answer to above questions, the actual development of science in the human societies, its grammar, culture and history. Only then a true sense of its claims and raison d'être to which it is demanding such a forceful attention can be ascertained. This becomes more obvious, when one looks at the theoretical/conceptual apparatus employed by the scientist in the description of physical reality, and which is used for the production of scientific knowledge. It does not happen in ivory-towers, the formulation of hypothesis, construction of theories, tools of observation, experimental devices, testing and acceptance of concepts, and discoveries of new laws and phenomena are governed by the prevalent rules of the game, by doing "normal science" by prejudices, by propaganda, by tactics, and by intimation. This situation is further constrained and escalated by the profound cosmological as well as ideological idiosyncrasies of its practitioners. All those conditions collectively influence the growth and definition of scientific knowledge, in short one has to research the historical grammar of scientific knowledge in action, for understanding it's nature and structure, as practiced by its professors and not by the terms, rules of logician or methodologist. Therefore, any methodology of scientific knowledge which overlooks the cultural-historical context of or what Feyerabend aptly terms, as the "cosmology" underlying the scientist's theoretical constructions, and conversely imposing rules for doing good science or laying down criteria for what is scientific and what is not, losses touch with the culture of scientist; and the larger culture of humanity in which he is situated. It reduces science "the great system of utterances which try to evoke and impose correct mode of feelings" on the scientist's mind and his activity (scientific inquiry) to mere logic chopping, and conceptual purity, which is meaningless and devoid of social significance. Described in strict rules (and rationalities) scientific knowledge, assumes a confrontational posture towards other modes of knowledge acquisition or else denigrate them. This is what has created deep cleavages in the cultural life of modern man. Why? The answer lies in technology; which is a facade of modern scientific knowledge and has become so powerful a business in the hands of its practitioners and the institutions which support the entire enterprise in the name of R and D. That is why Feyerabend, in his harsh polemical style dub this professional business class of scientists as "human ants who excel in the solution of tiny problems but who cannot make sense of anything transcending their domain of competence". Simultaneously, this seems pertinent to remark that the teaching of scientific theories as a factual cosmology of the time, not to be questioned, challenged or replaced by other (different) cosmologies (which possesses its own rationality and history) and its enforcement by powerful state apparatus, media and popular propaganda, constitute a Medieval attitude of blind following of an infallible authority (which is perhaps vice versa). This attitude is to be overthrown if scientific knowledge has to become a cherished asset to other domains of humanity's culture. Liberation of human society from this business-like and infallible attitude of science and scientist, philosophy must come forward to create a sense of commitment and participation in scientist's view of the world, so as to facilitate a humanistic growth of scientific knowledge.

From Newton to Prigogine, science (and scientific knowledge) has traveled a long treacherous path to man's self-knowledge and discovery of the languages which universe speak, and is somewhat similar to Mao-Tse-Tung's political Long March in 1930's. The classical notions of the world have been transformed altogether into new conceptual frameworks. A universe which was a symbol of peaceful automation, timeless, orderly, obeying mathematical laws, now appears to be after basic discoveries in quantum mechanics, thermodynamics and bio-sciences, in a state of flux, multiplicity, complexity and temporality, trapped in fluctuations, disturbances, disequilibrium and time-oriented activity. Time creates us and the new structures of cosmos. It is no more a clock work type of reality, static and determined. The role of passions, emotions and participation in the act of scientific discovery has now firmly been established "A scientific theory is akin to a work of art".

Scientific knowledge is no more superior to other modes of knowledge acquisition rather it should complement other domains of human experience such as myth and religious experience. It is "one of the many instruments man has invented to cope with his surrounding" and consequently not infallible. Diachotomic consciousness of 18th and 19th century scientific world view has been replaced by a sense of being part and parcel of this world and its eventual destiny. This feeling of involvement in the life of cosmos was intuitively captured by Mir Taqui Mir, a classical Urdu poet of 18th century Mughal India in one of his couplet. He goes on to say:

> Breathe in slowly, this mirror-like cosmos
> Is a work delicate and subtle?

Chapter 2

PHENOMENA OF THE UNIVERSE AND NATURAL HISTORY FOR THE BUILDING UP OF PHILOSOPHY

Francis Bacon

Since it seems to me that people do not keep strictly to the straight and narrow when forming their opinions or putting things to the test, I have decided to use all the means at my disposal to remedy this misfortune. For in nothing else does the aspiration to deserve well show itself than it things are so arranged that people, freed both from the hobgoblins of belief and blindness of experiments, may enter into a more reliable and sound partnership with things by, as it were, a certain literate experience. For in this way the intellect is both set up in safety and in its best state, and it will besides be at the ready and then come upon harvests of useful things.

Now the beginnings of this enterprise must in general be drawn from natural history; for the whole body of Greek philosophy with its sects of all kinds, and all the other philosophy we possess seem to me to be founded on too narrow a natural-historical basis, and thus to have delivered its conclusions on the authority of fewer data than was appropriate. For having snatched certain things from experience and tradition, things sometimes not carefully examined or ideas nor securely established, they leave the rest to meditation and intellectual agitation, employing *Dialectic* inspire greater confidence in the matter.

But the chemists and the whole pack of mechanics and empirics, should they have the temerity to attempt contemplation and philosophy, being accustomed to meticulous subtlety in a few things, they twist by extraordinary means all the rest into conformity with them and promote opinions more odious and unnatural than those advanced by the very rationalists. For the latter take for the matter of philosophy very little out of many things, the former a great deal out of a few, but in truth those courses are weak and past cure. But the Natural History which has been accumulated hitherto may seem abundant on casual inspection, while in reality it is sketchy and useless, and not even of the kind I am seeking. It has not been stripped of fables and ravings, and it rushes into antiquity, philology and superfluous narratives, neglectful and high-handed in matters of weight, overscrupulous and immoderate in matters of no importance. But the worst thing about this abundance is that it has embraced the inquiry into things natural but largely spurned that into things mechanical. Now the latter are far better than the former for examining nature's recesses; for nature of its own accord, free and shifting, disperses the intellect and confuses it with its variety, but in mechanical operations the judgment is concentrated, and we see nature's modes and processes, not just its effects. Yet, on the other hand, all the subtlety of mechanics stops short of what I am seeking. For the craftsman, intent on his work and its end, does not direct his mind or put his hand to other things, things which perhaps do more for the inquiry into nature.

Therefore we need more meticulous care and handpicked trials, not to mention funding and the utmost patience besides. For it has ruined everything in the experimental field that right from the beginning men have continually aimed at Experiments of Fruit not ones of Light, and have devoted their energies entirely to producing some splendid work, not to revealing nature's oracles, which is the work of works and encompasses in itself all power. It also comes about from men's misguided conceit that they have mostly applied themselves to things hidden and rare, and put their efforts and inquiry into those while spurning common experiments and observations, and this seems to have come about either because they sought admiration and fame, or because they fell for the belief that the function of philosophy lies in accommodating and reducing rarer events to those which occur familiarly, not equally to unearthing the causes of these common things themselves and deeper causes of those causes.

But the main point of the whole accusation against natural history is that men have gone astray not only in the work, but in its very plan. For the natural history which is in existence seems to have been composed either for the usefulness of the experiments themselves, or for the agreeableness of their narratives, and to have been made for their own sake, not so as to furnish the makings of philosophy and the sciences and as it were breast-feed them.

Thus, as far as it is within my power, I do not wish to fail to do my duty in this matter. For I have long since decided how much I should grant to abstract philosophies. Indeed, I believe that I hold fast to the ways of true and good induction, in which all things lie, and which can help the frail and crippled faculty of human intellect towards the sciences, as by mechanical aids or by some thread to guide it through a labyrinth. Nor am I unaware that if I had been willing to restrict that instauration of the sciences which I have in mind to any of the greater inventions, I could perhaps have harvested a greater crop of honor. But since God has given me a mind which knows how to submit itself to things and which readily rejects the specious out of a sense of what is right and from confidence that things will turn out well, I have also taken upon myself that part of the work which I think others have wanted either to avoid entirely, or to treat in a way different from my idea of it.

But there are two things which I wish to warn people about in this connection both for the future and, since I am girding myself for the very thing itself, for now especially. The *first* is to get rid of that idea which, though it be utterly false and harmful, easily invades and takes hold of men's minds, namely that the inquiry into particulars is something infinite and without end, when it would be truer to say that the way of opinions and disputations is the trifling one; but in fact these vain imaginings are condemned to perpetual errors and infinite disturbances, whereas particulars and the information of the sense (which, when individuals and the gradations of things have been left out, is sufficient for the inquiry into truth) allow understanding for certain, and that, to be sure, neither forlorn nor hopeless.

The *second* is that I would have men never forget what is involved and, when they have come across troops of thoroughly vulgar things, things slight and to all appearances frivolous, even vile, and which (as the man says) must be brought in with an apology, they do not think I am trifling, or reducing the human mind to things beneath its dignity. For these things are neither examined nor described for their own sake, but in fact there is simply no other alternative open to the human intellect, and the grounds of the work are left insecure without them. I am then certainly undertaking the most serious business of all and most worthy of the human mind, that nature's light, pure and quite unclouded by vain imagination (that light whose name has sometimes been mentioned thus far, while people have known nothing about the thing itself), may be lit in this age of ours by a torch furnished and brought near by the Divine Will.

For I do not hide the fact that I believe that preposterous subtlety of argument and thought can by no means put things right again, though all the intellects of all ages be gathered together, when, at the proper time, the subtlety and truth of the basic information or true induction have been overlooked or incorrectly established, but that nature, like fortune, is long-haired at the front and bald at the back. It remains, therefore, for the matter to be attempted anew, and that with better help and with the zeal of opinions laid aside, so that we may enter into the kingdom of philosophy and the sciences (in which human power is situated, for nature is conquered only by obeying it) in the way that we gain access to the Kingdom of Heaven, which none may enter save in the likeness of a little child. Yet I do not wholly despise the base and indiscriminate custom of working by experiments themselves (for it may doubtless suggest very many useful things to men's knowledge and invention, according to the variety of their arts and capacities), nevertheless I think it is something very trivial in comparison with that entrance into human knowledge and power which I hope for from the Divine Mercy, which indeed I again humbly beseech to allow me to endow the human family with new alms through my efforts.

The nature of things is either *free*, as in *species*, or *disturbed*, as in *monsters*, or *confined*, as in *experiments* of the Arts; yet its deeds of whatever kind are worthy of report and history. But the *History of Species* currently available, as for example of plants, animals, metals and fossils, is puffed up and full of curiosities; the *History of Marvels* empty and based on rumor; the *History of Experiments* detective, attempted piecemeal, dealt with carelessly, and entirely for practical not philosophical use.

Therefore it is my resolve to curb the History of Species, to shake our and purify the History of Marvels, but to our special effort into Mechanical and Artificial Experiments where nature gives in to human intervention. For what are the sports and frivolities of nature to us? That is, the tiny differences of species as to shape, which contribute nothing to works but in which Natural History none the less abounds. Now knowledge of Marvels certainly pleases me, if it be purified and sifted; but why in the final analysis is it pleasing? Not for the fun of being astonished, but because it often reminds Art of its duty to lead nature knowingly where it has itself sometimes gone before of its own accord.

In general I assign the leading roles in shedding light on nature to artificial things, not only because they are most useful in themselves, but because they are the most trustworthy interpreters of natural things. Can it be said that anyone had just happened to explain the nature of lightning or a rainbow as clearly before the principles of each had been demonstrated by artillery or the artificial simulacra of rainbows on a wall? But if they are trustworthy interpreters of causes, they will also be sure and fertile indicators of effects and of works. However, I do not think it appropriate to divide my history in accordance with this threefold partition, so as to deal with singular instances separately, but I shall mix the three kinds, joining things natural with artificial. The ordinary as well as extraordinary pays very close attention to all the most useful ones.

Now it would be more usual to begin with the phenomena of the ether. But I, sacrificing nothing of the seriousness of my undertaking, shall give priority to things which make up and answer to a nature more general, in which both globes share. I shall begin in fact with a history of bodies according to the difference which seems the simplest, that is, the abundance or paucity of the matter contained and spread out within the same space or boundaries, seeing indeed that none of the pronouncements about nature is truer than that double proposition. *Nothing comes from nothing, nor is anything reduced to nothing*, but *the very quantum of nature*, or *the whole sum of matter* always remains and stays the same, and is in no way increased or diminished. Moreover, it is no less certain. even though not so clearly noted or asserted (whatever stories people make up about the impartial potential of matter towards forms) that more or less of this quantity of matter is contained in the same volumes of space according to the diversity of the bodies which occupy them, bodies some of which we find to be very obviously more compact, others more extended or diffuse. For a vessel or cauldron filled with water and air does not hold an equal portion of matter, but more of the one and less of the other. Therefore if someone claimed that a given amount of water could be made from the same amount of air, it would be the same as saying that something can come from nothing. For what you deem to be lacking from the quantity of matter would have to have been made up from nothing. On the other hand, if someone claimed that a given amount of water could be turned into the same amount of air, it would be the same as saying that something can be reduced to nothing. For what you deem to be extra in the quantity of matter would likewise have to have vanished into nothingness. There is no doubt in my mind that this business is capable of being reduced to calculation, to indefinite proportions perhaps in some things, but to ones precise and certain in others, and known to nature. As, for example, if someone said that the concentration of matter in a body of gold exceeded that of a body of spirit of wine by a factor of twenty to one or thereabouts, he would not be wrong. So as I now mean to present the history I mentioned concerning the abundance and paucity of matter, and its coming together and expansion, things from which the notions of *Dense* and *Rare* (if properly understood) take their origin, I shall so order matters that I shall draw up the relative figures for different bodies (as of gold, water, oil, air and flame) first. Then after examining these, I shall record with calculations or ratios the retreats and expatiations of each particular body. For a given body, even without anything being added to it or taken away, or at least nor in proportion to its contraction and extension, allows itself to be gathered by various impulses both external and internal into a larger or smaller sphere. Sometimes the body struggles and strives to restore itself into its old sphere, sometimes it clearly goes beyond that and does not try to revert. Here I shall first record the courses, differences and proportions of any natural body (as to its extent) compared with its openings and closings up, that is, with its powders, its calices, its verifications, its dissolutions, its distillations, vapors and breaths, its exhalations and inflammations; then I shall set out the actions and motions themselves, the progressions and the limits of contraction and dilatation, and when bodies restore themselves and when they go beyond than in respect of their extent; but I shall especially note the efficient causes and media by means of which such contractions and dilatations of bodies come about; and meanwhile I shall in passing append the virtues and actions which bodies get and take on from such compressions and dilatations.

And since I know well how difficult a thing it is, in the present climate of opinion, to familiarize oneself with nature right from the very beginning, I shall add my own observations to gain men's attention and arouse them to contemplation. Now as far as the demonstration or revealing of the density and rarity of bodies is concerned, I have no doubt or hesitation that as to *dense* and *palpable bodies the motion of gravity* (as they call it) may be taken as the best and most ready test, for the more compact the body, the heavier it is. But when it comes to the level of airy and spiritual things, then scales will for sure be of no use to me, and I shall need another kind of industry. I shall begin, however, with *Gold*: which of all the things we have (for philosophy has not grown up enough for us to say anything for certain about the bowels of the Earth) is the heaviest and contains the most matter in the smallest space, and I shall relate the ratios of the rest to the sphere of this body, with the reminder that I am not dealing here with the history of weights except in so far as it sheds light for demonstrating the space or dimensions of bodies ...

Chapter 3

DISCOURSE ON THE METHOD OF RIGHTLY CONDUCTING THE REASON AND SEEKING FOR TRUTH IN THE SCIENCES

Rene Descartes

If this Discourse appears too long to be read all at once, it may be separated into six portions. And in the first there will be found various considerations respecting the sciences; in the second the principal rules regarding the Method which the author has sought out; while in the third are some of the rules of morality which he has derived from this Method. In the fourth are the reasons by which he proves the existence of God and of the human soul which form the foundation of his Metaphysic. In the fifth, the order of the questions regarding physics which he has investigated, and particularly the explanation of the movement of the heart, and of some other difficulties which pertain to medicine as also the difference between the soul of man and that of the brutes. And in the last part the questions raised relate to those matters which the author believes to be requisite in order to advance further in the investigation of nature, in addition to the reasons that caused him to write.

PART I

Good sense is of all things in the world the most equally distributed, for everybody thinks himself so abundantly provided with it, that even those most difficult to please in all other matters do not commonly desire more of it than they already possess. It is unlikely that this is an error on their part; it seems rather to be evidence in support of the view that the power of forming a good judgment and of distinguishing the true from the false, which is properly speaking what is called Good sense or Reason, is by nature equal in all men. Hence too it will show that the diversity of our opinions does not proceed from some men being more rational than others, but solely from the fact that our thoughts pass through diverse channels and the same objects are not considered by all. For to be possessed of good mental powers is not sufficient; the principal matter is to apply them well. The greatest minds are capable of the greatest vices as well as of the greatest virtues and those who proceed very slowly may, provided they always follow the straight road, really advance much faster than those who, though they run, forsake it.

For myself I have never ventured to presume that my mind was in any way more perfect than that of the ordinary man; I have even longed to possess thought as quick;, or an imagination as accurate and distinct, or a memory as comprehensive or ready, as some others. And besides these I do not know any other qualities that make for the perfection of the human mind. For as to reason or sense, inasmuch as it is the only thing that constitutes us men and distinguishes us from the brutes, I would fain believe that it is to be found complete in each individual, and in this I follow the common opinion of the philosophers, who say that the question of more or less occurs only in the sphere of the *accidents* and does not affect the *forms* or natures of the *individuals* in the same *species*.

But I shall not hesitate to say that I have had great good fortune from my youth up, in lighting upon and pursuing certain paths which have conducted me to considerations and maxims from which I have formed a Method, by whose assistance it appears to me I have the means of gradually increasing my knowledge and of little by little raising it to the highest possible point which the mediocrity of my talents and the brief duration of my life can permit me to reach. For I have already reaped from it fruits of such a nature that, even though I always try in the judgments I make on myself to lean to the side of self-depreciation rather than to that of arrogance, and though, looking with the eye of a philosopher on the diverse actions and enterprises of all mankind, I find scarcely any which do not seem to me vain and useless, I do not cease to receive extreme satisfaction in the progress which I seem to have already made in the search after truth, and to form such hopes for the future as to venture to believe that, if amongst the occupations of men, simply as men, that is the one in particular that is excellent and important, that is the one which I have selected.

It must always be recollected, however, that possibly I deceive myself, and that what I take to be gold and diamonds is perhaps no more than copper and glass. I know how subject we are to delusion in whatever touches ourselves, and also how much the judgments of our friends ought to be suspected when they are in our favor. But in this Discourse I shall be very happy to show the Paths I have followed, and to set forth my life as in a picture, so that everyone may judge of it for himself; and thus in learning from the common talk what are the opinions which are held of it, a new means of obtaining self-instruction will be reached, which I shall add to those which I have been in the habit of using.

Thus my design is not here to teach the Method which everyone should follow in order to promote the good conduct of his Reason, but only to show in what manner I have endeavored to conduct my own. Those who set about giving precepts must esteem themselves more skilful than those to whom they advance them, and if they fall short in the smallest matter they must of course take the blame for it. But regarding this Treatise simply as a history, or, if you prefer it, a fable in which, amongst certain things which may be imitated, there are possibly others also which it would not be right to follow, I hope that it will be of use to some without being hurtful to any, and that all will thank me for my frankness.

I have been nourished on letters since my childhood, and since I was given to believe that by their means a clear and certain knowledge could be obtained of all that is useful in life, I had an extreme desire to acquire instruction. But as soon as I had achieved the entire course of study at the loss of which one is usually received into the ranks of the learned, I entirely changed my opinion. For I found myself embarrassed with so many doubts and errors that it seemed to me that the effort to instruct myself had no effect other than the increasing discovery of my own ignorance. And yet I was studying at one of the most celebrated Schools in Europe, where I thought that there must be men of learning if they were to be found anywhere in the world. I learned there all that others learned; and not being satisfied with the sciences that we were taught, I even read through all the books which fell into my hands, treating of what is considered most curious and rare. Along with this I knew the judgments that others had formed of me, and I did not feel that I was esteemed inferior to my fellow-students, although there were amongst them some destined to fill the places of our masters. And finally our century seemed to me as flourishing, and as fertile in great minds, as any which had proceeded. And this made me take the liberty of judging all others by myself and of coming to the conclusion that there was no learning in the world such as I was formerly led to believe it to be.

I did not omit, however, always to hold in esteem those exercises which are the occupation of the Schools. I knew that the Languages which one learns there are essential for the understanding of all ancient literature; that fables with their charm stimulate the mind and histories of memorable deeds exalt it; and that, when read with discretion, these books assist in forming a sound judgment I was aware that the reading of all good books is indeed like a conversation with the noblest men of past centuries who were the authors of them, nay a carefully studied conversation, in which they reveal to us none but the best of their thoughts. I deemed Eloquence to have a power and beauty beyond compare; that Poesy has most ravishing delicacy and sweetness; that in Mathematics there are the subtlest discoveries and inventions which may accomplish much, both in satisfying the curious, and in furthering all the arts, and in diminishing man's labor; that those writings that deal with Morals contain much that is instructive, and many exhortations to virtue which are most useful; that Theology points out the way to heaven; that Philosophy teaches us to speak with an appearance of truth on all things, and causes us to be admired by the less learned; than Jurisprudence, Medicine and all other sciences bring honor and riches to those who cultivate them; and finally that it is good to have examined all things, even those most full of superstition and falsehood, in order that we may know their just value, and avoid being deceived by them.

But I considered that I had already given sufficient time to languages and likewise even to the reading of the literature of the ancients, both their histories and their fables. The conversation with those of other centuries is almost the same thing as to travel. It is good to know something of the customs of different peoples in order to judge more sanely of our own, and not to think that everything of a fashion not ours is absurd and contrary to reason, as who those who have seen nothing. But when one employs too much time in travelling, one becomes a stranger in one's own country, and when one is too curious about things which were practiced in past centuries, one is usually very ignorant about those which are practiced in our own time. Besides, fables make one imagine many events possible which in reality are not so, and even the most accurate of histories, if they do not exactly misrepresent or exaggerate the value of things in order to render them more worthy of being read, at least omit in them all the circumstances which are basest and least notable; and from this fact it follows that what is retained is not portrayed as it really is and that those who regulate their conduct by examples which they derive from such a source, are liable to fall into the extravagances of the knights-errant of Romance, and form projects beyond their power of performance.

I esteemed Eloquence most highly and I was enamored of Poesy, but I thought that both were gifts of the mind rather than fruits of study. Those who have the strongest power of reasoning, and who most skillfully arrange their thoughts in order to tender them clear and intelligible, have the best power of persuasion even if they can but speak the language of Lower Brittany and have never learned Rhetoric. And those who have the most delightful original ideas and who know how to express them with the maximum of style and suavity, would not fail to be the best poets even if the art of Poetry were unknown to them.

Most of all was I delighted with Mathematics because of the certainty of its demonstrations and the evidence of its reasoning; but I did not yet understand its true use, and, believing that it was of service only in the mechanical arts, I was astonished that, seeing how firm and solid was its basis, no loftier edifice had been reared thereupon. On the other hand I compared the works of the ancient pagans which deal with Morals to palaces most superb and magnificent, which are yet built on sand and mud alone. They praise the virtues most highly and show them to be more worthy of being prized than anything else in the world, but they do not sufficiently teach us to become acquainted with them, and often that which is called by a fine name is nothing but insensibility, or pride, or despair, or parricide.

I honoured our theology and aspired as much as anyone to reach to heaven, but having learned to regard it as a most highly assured fact that the road is not less open to the most ignorant than to the most learned, and that the revealed truths which conduct thither are quite above our intelligence, I should not have dared to submit them to the feebleness of my reasoning; and I thought that, in order to undertake to examine them and succeed in so doing, it was necessary to have some extraordinary assistance from above and to be more than a mere man.

I shall not say anything about Philosophy, but that, seeing that it has been cultivated for many centuries by the best minds that have ever lived, and that nevertheless no single thing is to be found in it which is not subject of dispute, and in consequence which is not dubious I had not enough presumption to hope to fare better there than other men had done. And also, considering how many conflicting opinions there may be regarding the self-same matter, all supported by learned people, while there can never be more than one which is true, I esteemed as well-nigh false all that only went as far as being probable.

Then as to the other sciences, inasmuch as they derive their principles from Philosophy, I judged that one could have built nothing solid on foundations so far from firm. And neither the honor nor the promised gain was sufficient to persuade me to cultivate them, for, thanks be to God, I did not find myself in a condition which obliged me to make a merchandise of science for the improvement of my fortune; and, although I did not pretend to scorn all glory like the Cynics, I yet had very small esteem for what I could not hope to acquire, excepting through fictitious tittles. And, finally, as to false doctrines, I thought that I already knew well enough what they were worth to be subject to deception neither by the promises of an alchemist, the predictions of an astrologer, the impostures of a magician, the artifices or the empty boastings of any of those who make a profession of knowing that of which they are ignorant.

This is why, as soon as age permitted me to emerge from the control of my tutors, I entirely quitted the study of letters. And resolving to seek no other science than that which could be found in myself, or at least in the great book of the world, I employed the rest of my youth in travel, in seeing courts and armies, in intercourse with men of diverse temperaments and conditions, in collecting varied experiences, in proving myself in the various predicaments in which I was placed by fortune, and under all circumstances bringing my mind to bear on the things which came before it, so that I might derive some profit from my experience. For it seemed to me that I might meet with much more truth in the reasoning that each man makes on the matters that specially concern him, and the issue of which would very soon punish him if he made a wrong judgment, than in the case of those made by a man of letters in his study touching speculations which lead to no results and which bring about no other consequences to himself excepting that he will be all the more vain the more they are removed from common sense, since in this case it proves him to have employed so much the more ingenuity and skill in trying to make them seem probable. And I always had an excessive desire to learn to distinguish the true from the false, in order to see clearly in my actions and to walk with confidence in this life.

It is true that while I only considered the manners of other men I found in them nothing to give me settled convictions; and I remarked in them almost as much diversity as I had formerly seen in the opinions of philosophers. So much was this the case that the greatest profit which I derived from their study was that, in seeing many things which, although they seem to us very extravagant and ridiculous, were yet commonly received and approved by other great nations, I learned to believe nothing too certainly of which I had only been convinced by example and custom. Thus, little by little I was delivered from many errors which might have obscured our natural vision and rendered us less capable of listening to Reason. But after I had employed several years in thus studying the book of the world and trying to acquire some experience, I one day formed the resolution of also making me an object of study and of employing all the strength of my mind in choosing the road I should follow. This succeeded much better; it appeared to me, than if I had never departed either from my country or my books.

PART II

I was then in Germany, to which country I had been attracted by the wars which are not yet at an end. And as I was returning from the coronation of the Emperor to join the army, the setting in of winter detained me in a quarter where, since I found no society to divert me, while fortunately I had also no cares or passions to trouble me, I remained the whole day shut up alone in a stove-heated room, where I had complete leisure to occupy myself with my own thoughts. One of the first of the considerations that occurred to me was that there is very often less perfection in works composed of several portions, and carried out by the hands of various masters, than in those on which one individual alone has worked. Thus we see that buildings planned and carried out by one architect alone are usually more beautiful and better proportioned than those which many have tried to put in order and improve making use of old walls which were built with other ends in view. In the same way also, those ancient cities which, originally mere villages, have become in the process of time great towns, are usually badly constructed in comparison with those which are regularly laid out on a plain by a surveyor who is free to follow his own ideas. Even though, considering their buildings each one apart, there is often as much or more display of skill in the one case than in the other, the former have large buildings and small buildings indiscriminately placed together, thus rendering the streets crooked and irregular, so that it might be said that it was chance rather than the skill of men guided by reason that led to such an arrangement. And if we consider that this happens despite the fact that from all time there have been certain officials who have had the special duty of looking after the buildings of private individuals in order that they may be public ornaments, we shall understand how difficult it is to bring about much that is satisfactory in operating only upon the works of others. Thus I imagined that those people who were once half-savage, and who have become civilized only by slow degrees, merely forming their laws as the disagreeable necessities of their crimes and quarrels constrained them, could not succeed in establishing so good a system of government as those who, from the time they first came together as communities, carried into effect the constitution laid down by some prudent legislator. Thus it is quite certain that the constitution of the true Religion whose ordinances are of God alone is incomparably better regulated than any other. And, to come down to human affairs, I believe that if Sparta was very flourishing in former times, this was not because of the excellence of each and every one of its laws, seeing that many were very strange and even contrary to good morals, but because, being drawn up by one individual, they all tended towards the same end. And similarly I thought that the sciences found in books - in those at least whose reasoning are only probable and which have no demonstrations, composed as they are of the gradually accumulated opinions of many different individuals - do not approach so near to the truth as the simple reasoning which a man of common sense can quite naturally carry out respecting the things which come immediately before him. Again I thought that since we have all been children before being men, and since it has for long fallen to us to be governed by our appetites and by our teachers who often enough contradicted one another, and none of whom perhaps counseled us always for the best), it is almost impossible that our judgments should be so excellent or solid as they should have been had we had complete use of our reason since our birth, and had we been guided by its means alone.

It is true that we do not find that all the houses in a town are raised to the ground for the sole reason that the town is to be rebuilt in another fashion, with streets made more beautiful; but at the same time we see that many people cause their own houses to be knocked down in order to rebuild them, and that sometimes they are forced so to do where there is danger of the houses falling of themselves, and when the foundations are not secure. From such examples I argued to myself that there was no plausibility in the claim of any private individual to reform a state by altering everything, and by overturning it throughout, in order to set it right again. Nor is it likewise probable that the whole body of the Sciences, or the order of teaching established by the Schools, should be reformed. that as regards all the opinions which up to this time I had embraced, I thought I could not do better than endeavor once for all to sweep them completely away, so that they might later on be replaced, either by others which were better, or by the same, when I had made them conform to the uniformity of a rational scheme And I firmly believed that by this means I should succeed in directing my life much better than if I had only built on old foundations, and relied on principles of which I allowed myself to be in youth persuaded without having inquired into their truth. For although in so doing I recognized various difficulties, these were at the same time not insurmountable, nor comparable to those which are found in reformation of the most insignificant kind in matters which concern the public. In the case of great bodies it is too difficult a task to raise them again when they are once thrown down or even to keep them in their places when once thoroughly shaken; and their fall cannot be otherwise than very violent. Then as to any imperfections that they may possess (and the very diversity that is found between them is sufficient to tell us that these in many cases exist) custom has doubtless greatly mitigated them, while it has also helped us to avoid, or insensibly corrected a number against which mere foresight would have found it difficult to guard. And finally the imperfections are almost always more supportable than would be the process of removing them, just as the great roads which wind about amongst the mountains become, because of being frequented, little by little so well beaten and easy that it is much better to follow them than to try to go more directly by climbing over rocks and descending to the foot of precipices.

This is the reason why I cannot in any way approve of those turbulent and unstable spirits who, being called neither by birth nor fortune to the management of public affairs, never fail to have always in their minds some new reforms. And if I thought that in this treatise there was contained the smallest justification for this folly, I should be very sorry to allow it to be published. My design thus never extended beyond trying to reform my own opinion and to build on a foundation which is entirely my own. If my work has given me a certain satisfaction, so that I here present to you a draft of it, I do not so do because I wish to advise anybody to imitate it. Those to whom God has been most beneficent in the bestowal of His graces will perhaps form designs which are more elevated; but I fear much that this particular one will seem too venturesome for many. The simple resolve to strip myself of all opinions and beliefs formerly received is not to be regarded as an example that each man should follow, and the world may be said to be mainly composed of those classes of minds neither of which could prudently adopt it. There are those who believing themselves to be cleverer than they are, cannot restrain themselves from being precipitate in judgment and have not sufficient patience to arrange their thoughts in proper order; hence, once a man of this description had taken the liberty of doubting the principles he formerly accepted, and had deviated from the beaten track, he would never be able to maintain the path which must be followed to reach the appointed end more quickly, and he would hence remain wandering astray all through his life. Secondly, there are those who having reason or modesty enough to judge that they are less capable of distinguishing truth from falsehood than some others from whom instruction might be obtained, are right in contenting themselves with following the opinions of these others rather than in searching better ones for themselves.

For myself I should doubtless have been of these last if I had never had more than a single master, or had I never known the diversities which have from all time existed between the opinions of men of the greatest learning. But I had been taught, even in my college days, that there is nothing imaginable so strange or so little credible that it has not been maintained by one philosopher or other, and I further recognized in the course of my travels that all those whose sentiments are very contrary to ours are yet not necessarily barbarians or savages, but may be possessed of reason in as great or even a greater degree than ourselves. I also considered how very different the self-same man, identical in mind and spirit, may become, according as he is brought up from childhood amongst the French or Germans, or has passed his whole life amongst Chinese or cannibals. I likewise noticed how even in the fashions of one's clothing the same thing that pleased us ten years ago, and which will perhaps please us once again before ten years are passed, seems at the present time extravagant and ridiculous. I thus concluded that it is much more custom and example that persuade us than any certain knowledge, and yet in spite of this the voice of the majority does not afford a proof of any value in truths a little difficult to discover, because such truths are much more likely to have been discovered by one man than by a nation. I could not, however, put my finger on a single person whose opinions seemed preferable to those of others, and I found that I was, so to speak, constrained myself to undertake the direction of my procedure.

But like one who walks alone and in the twilight I resolved to go so slowly, and to use so much circumspection in all things, that if my advance was but very small, at least I guarded myself well from falling. I did not wish to set about the final rejection of any single opinion which might formerly have crept into my beliefs without having been introduced there by means of Reason, until I had first of all employed sufficient time in planning out the task which I had undertaken, and in seeking the true Method of arriving at a knowledge of all the things of which my mind was capable.

Among the different branches of Philosophy, I had in my younger days to a certain extent studied Logic; and in those of Mathematics, Geometrical Analysis and Algebra - three arts or sciences which seemed as though they ought to contribute something to the design I had in view. But in examining them I observed in respect to Logic that the syllogisms and the greater part of the other teaching served better in explaining to others those things that one knows (or like the art of Lully, in enabling one to speak without judgment of those things of which one is ignorant) than in learning what is new. And although in reality Logic contains many precepts which are very true and very good, there are at the same time mingled with them so many others which are hurtful or superfluous that it is almost as difficult to separate the two as to draw a Diana or a Minerva out of a block of marble which is not yet roughly hewn. And as to the Analysis of the ancients and the Algebra of the moderns, besides the fact that they embrace only matters the most abstract, such as appear to have no actual use, the former is always so restricted to the consideration of symbols that it cannot exercise the Understanding without greatly fatiguing the Imagination; and in the latter one is so subjected to certain rules and formulas that the result is the construction of an art which is confused and obscure, and which embarrasses the mind, instead of a science which contributes to its cultivation. These made me feel that some other Method must be found, which, comprising the advantages of the three, is yet exempt from their faults. And as a multiplicity of laws often furnishes excuses for evil-doing, and as a State is hence much better ruled when, having but very few laws, these are most strictly observed; so, instead of the great number of precepts of which Logic is composed, I believed that I should find the four which I shall state quite sufficient, provided that I adhered to a firm and constant resolve never on any single occasion to fail in their observance.

The first of these was to accept nothing as true which I did not clearly recognize to be so: that is to say, carefully to avoid precipitation and prejudice in judgments and to accept in them nothing more than what was presented to my mind so clearly and distinctly that I could have no occasion to doubt it.

The second was to divide up each of the difficulties which I examined into as many parts as possible, and as seemed requisite in order that it might be resolved in the best manner possible.

The third was to carry on my reflections in due order, commencing with objects that were the most simple and easy to understand, in order to rise little by little, or by degrees, to knowledge of the most complex, assuming an order, even if a fictitious one, among those which do not follow a natural sequence relatively to one another.

The last was in all cases to make enumerations so complete and reviews so general that I should be certain of having omitted nothing.

Those long chains of reasoning, simple and easy as they are, of which geometricians make use in order to arrive at the most difficult demonstrations, had caused me to imagine that all those things which fall under the cognizance of man might very likely be mutually related in the same fashion; and that, provided only that we abstain from receiving anything as true which is not so, and always retain the order which is necessary in order to deduce the one conclusion from the other, there can be nothing so remote that we cannot reach to it, nor so recondite that we cannot discover it. And I had not much trouble in discovering which objects it was necessary to begin with, for I already knew that it was with the most simple and those most easy to apprehend. Considering also that of all those who have hitherto sought for the truth in the Sciences, it has been the mathematicians alone who have been able to succeed in making any demonstrations, that is to say producing reasons which are evident and certain, I did not doubt that it had been by means of a similar kind that they carried on their investigations. I did not at the same time hope for any practical result in so doing, except that my mind would become accustomed to the nourishment of truth and would not content itself with false reasoning. But for all that I had no intention of trying to master all those particular sciences that receive in common the name of Mathematics; but observing that, although their objects are different, they do not fail to agree in this, that they take nothing under consideration but the various relationships or proportions which are present in these objects, I thought that it would be better if I only examined these proportions in their general aspect, and without viewing them otherwise than in the objects which would serve most to facilitate a knowledge of them. Not that I should in any way restrict them to these objects, for I might later on all the more easily apply them to all other objects to which they were applicable. Then, carefully noted that in order to comprehend the proportions I should sometimes require to consider each one in particular, and sometimes merely keep them in mind, or take them in groups, I thought that, in order the better to consider them in detail, I should picture them in the form of lines, because I could find no method more simple nor more capable of being distinctly represented to my imagination and senses. I considered, however, that in order to keep them in my memory or to embrace several at once, it would be essential that I should explain them by means of certain formulas, the shorter the better. And for this purpose it was requisite that I should borrow all that is best in Geometrical Analysis and Algebra, and correct the errors of the one by the other.

As a matter of fact, I can venture to say that the exact observation of the few precepts which I had chosen gave me so much facility in sifting out all the questions embraced in these two sciences, that in the two or three months which I employed in examining them - commencing with the most simple and general, and making each truth that I discovered a rule for helping me to find others - not only did I arrive at the solution of many questions which I had hitherto regarded as most difficult, but, towards the end, it seemed to me that I was able to determine in the case of those of which I was still ignorant, by what means, and in how far, it was possible to solve them. In this I might perhaps appear to you to be very vain if you did not remember that having but one truth to discover in respect to each matter, whoever succeeds in finding it knows in its regard as much as can be known. It is the same as with a child, for instance, who has been instructed in Arithmetic and has made an addition according to the rule prescribed; he may be sure of having found as regards the sum of figures given to him all that the human mind can know. For, in conclusion, the Method which teaches us to follow the true order and enumerate exactly every term in the matter under investigation contains everything which gives certainty to the rules of Arithmetic.

But what pleased me most in this Method was that I was certain by its means of exercising my reason in all things, if not perfectly, at least as well as was in my power And besides this, I felt in making use of it that my mind gradually accustomed itself to conceive of its objects more accurately and distinctly; and not having restricted this Method to any particular matter, I promised myself to apply it as usefully to the difficulties of other sciences as I had done to those of Algebra. Not that on this account I dared undertake to examine just at once all those that might present themselves; for that would itself have been contrary to the order which the Method prescribes. But having noticed that the knowledge of these difficulties must be dependent on principles derived from Philosophy in which I yet found nothing to be certain, I thought that it was requisite above all to try to establish certainty in it. I considered also that since this endeavor is the most important in the entire world and that which precipitation and prejudice were most to be feared, I should not try to grapple with it till I had attained to a much riper age than that of three and twenty, which was the age I had reached. I thought, too, that I should first of all employ much time in preparing myself for the work by eradicating from my mind all the wrong opinions which I had up to this time accepted, and accumulating a variety of experiences fitted later on to afford matter for my reasoning, and by ever exercising myself in the Method which I had prescribed, in order more and more to fortify myself in the power of using it.

Part III

And finally, as it is not sufficient, before commencing to rebuild the house which we inhabit, to put it down and provide materials and an architect (or to act in this capacity ourselves, and make a careful drawing of its design), unless we have also provided ourselves with some other house where we can be comfortably lodged during the time of rebuilding, so in order that I should not remain irresolute in my notions while reason obliged me to be so in my judgments, and that I might not omit to carry on my life as happily as I could, I formed for myself a code of morals for the time being which did not consist of more than three or four maxims, which maxims I should like to enumerate to you.

The first was to obey the laws and customs of my country, adhering constantly to the religion in which by God's grace I had been instructed since my childhood, and in all other things directing my conduct by opinions the most moderate in nature, and the farthest removed from excess in all those which are commonly received and acted on by the most judicious of those with whom I might come in contact. For since I began to count my own opinions as naught, because I desired to place all under examination, I was convinced that I could not do better than follow those held by people on whose judgment reliance could be placed. And although such persons may possibly exist amongst the Persians and Chinese as well as amongst ourselves, it seemed to me that it was most expedient to bring my conduct into harmony with the ideas of those with whom I should have to live; and that, in order to ascertain that these were their real opinions, I should observe what they did rather than what they said, not only because in the corrupt state of our manners there are few people who desire to say all that they believe, but also because many are themselves ignorant of their beliefs. For since the act of thought by which we believe a thing is different from that by which we know that we believe it, the one often exists without the other. And amongst many opinions all equally received, I chose only the most moderate, both because these are always most suited for putting into practice, and probably the best (for all excess has a tendency to be bad), and also because I should have in a less degree turned aside from the right path, supposing that I was wrong, than if, having chosen an extreme course, I found that I had chosen amiss. I also made a point of counting as excess all the engagements by means of which we limit in some degree our liberty. Not that I hold in low esteem those laws which, in order to remedy the inconstancy of feeble souls, permit, when we have a good object in our view, that certain vows be taken, or contracts made, which oblige us to carry out that object. This sanction is even given for security in commerce where designs are wholly indifferent. But because I saw nothing in all the world remaining constant, and because for my own part I promised myself gradually to get my judgments to grow better and never to grow worse, I should have thought that I had committed a Serious ill against commonsense if, because I approved of something at one time, I was obliged to regard it similarly at a later time, after it had possibly ceased to meet my approval, or after I had ceased to regard it in a favorable light.

My second maxim was that of being as firm and resolute in my actions as I could be, and not to follow less faithfully opinions the most dubious, when my mind was once made up regarding them, than if these had been beyond doubt. In this I should be following the example of travelers, who, finding themselves lost in a forest, know that they ought not to wander first to one side and then to the other, nor, still less, to stop in one place, but understand that they should continue to walk as straight as they can in one direction, not diverging for any slight reason, even though it was possibly chance alone that first determined them in their choice. By this means if they do not go exactly where they wish, they will at least arrive somewhere at the end, where probably they will be better off than in the middle of a forest. And thus since often enough in the actions of life no delay is permissible, it is very certain that, when it is beyond our power to discern the opinions which carry most truth, we should follow the most probable; and even although we notice no greater probability in the one opinion than in the other, we at least should make up our minds to follow a particular one and afterwards consider it as no longer doubtful in its relationship to practice, but as very true and very certain, inasmuch as the reason which caused us to determine upon it is known to be so. And henceforward this principle was sufficient to deliver me from all the penitence and remorse which usually affect the mind and agitate the conscience of those weak and vacillating creatures who allow themselves to keep changing their procedure, and practice as good, things which they afterwards judge to be evil.

My third maxim was to try always to conquer myself rather than fortune, and to alter my desires rather than change the order of the world, and generally to accustom myself to believe that there is nothing entirely within our power but our own thoughts: so that after we have done our best in regard to the things that are without us, our ill-success cannot possibly be failure on our part. And this alone seemed to me sufficient to prevent my desiring anything in the future beyond what I could actually obtain, hence rendering me content; for since our will does not naturally induce us to desire nothing but what our understanding represents to it as in some way possible of attainment, it is certain that if we consider all good things which are outside of us as equally outside of our power, we should not have more regret in resigning those goods which appear to pertain to our birth, when we are deprived of them for no fault of our own, than we have in not possessing the kingdoms of China or Mexico. In the same way, in making what is called a virtue out of a necessity, we should no more desire to be well if ill, or free, if in prison, than we now do to have our bodies formed of a substance as little corruptible as diamonds, or to have wings to fly with like birds. I allow, however, that to accustom oneself to regard all things from this point of view requires long exercise and meditation often repeated; and I believe that it is principally ill this that is to be found the secret of those philosophers who, in ancient times, were able to free themselves from the empire of fortune, or, despite suffering or poverty, to rival their gods in their happiness. For, ceaselessly occupying themselves in considering the limits which were prescribed to them by nature, they persuaded themselves so completely that nothing was within their own power but their thoughts, that this conviction alone was sufficient to prevent their having any longing for other things. And they had so absolute a mastery over their thoughts that they had some reason for esteeming themselves as more rich and more powerful, and more free and more happy than other men, who, however favored by nature or fortune they might be, if devoid of this philosophy, never could arrive at all at which they aim.

And last of all, to conclude this moral code, I felt it incumbent on me to make a review of the various occupations of men in this life in order to try to choose out the best; and without wishing to say anything of the employment of others I thought that I could not do better than continue in the one in which I found myself engaged, that is to say, in occupying my whole life in cultivating Reason, and in advancing myself as much as possible in the knowledge of the truth in accordance with the method which I had prescribed myself. I had experienced so much satisfaction since beginning to use this method, that I did not believe that any sweeter or more innocent could in this life be found, - every day discovering by its means some truths which seemed to me sufficiently important, although commonly ignored by other men. The satisfaction which I had so filled my mind that all else seemed of no account, besides, the three preceding maxims were founded solely on the plan which I had formed of continuing to instruct myself. For since God has given to each of us some light with which to distinguish truth from error, I could not believe that I ought for a single moment to content myself with accepting the opinions held by others unless I had in view the employment of my own judgment in examining them at the proper time; and I could not have held myself free of scruple in following such opinions, if nevertheless I had not intended to lose no occasion of finding superior opinions, supposing them to exist; and finally, I should not have been able to restrain my desires nor to remain content, if I had not followed a road by which, thinking that I should be certain to be able to acquire all the knowledge of which I was capable, I also thought I should likewise be certain of obtaining all the best things which could ever come within my power. And inasmuch as our will impels us neither to follow after nor to flee from anything, excepting as our understanding represents it as good or evil, it is sufficient to judge wisely in order to act well, and the best judgment brings the best action - that is to say, the acquisition of all the virtues and all the other good things that it is possible to obtain. When one is certain that this point is reached, one cannot fail to be contented.

Having thus assured myself of these maxims, and having set them on one side along with the truths of religion which have always taken the first place in my creed, I judged that as far as the rest of my opinions were concerned; I could safely undertake to rid myself of them. And inasmuch as I hoped to hope to reach my end more successfully in converse with man than in living longer shut up in the warm room where these reflections had come to me I hardly awaited the end of winter before I once more set myself to travel. And in all the nine following years I did naught but roam hither and thither, trying to be a spectator rather than an actor in all the comedies the world displays. More especially did I reflect in each matter that came before me as to anything which could make it subject to suspicion or doubt, and give occasion for mistake, and I rooted out of my mind all the errors which might have formerly crept in. Not that indeed I imitated the skeptics, who only doubt for the sake of doubting, and pretend to be always uncertain; for, on the contrary, my design was only to provide myself with good ground for assurance, and to reject the quicksand and mud in order to find the rock or clay. In this task it seems to me, I succeeded pretty well, since in trying to discover the error or uncertainty of the propositions which I examined, not by feeble conjectures, but by clear and assured reasoning, I encountered nothing so dubious that I could not draw from it some conclusion that was tolerably secure, if this were no more than the inference that it contained in it nothing that vas certain. And just as in pulling down an old house we usually preserve the debris to serve in building up another, so ill destroying all those opinions which I considered to be ill-founded, I made various observations and acquired many experiences, which have since been of use to me in establishing those which are more certain. And more than this, I continued to exercise myself in the method which I had laid down for my use; for besides the fact that I was careful as a rule to conduct all my thoughts according to its maxims, I set aside some hours from time to time which I more especially employed in practicing myself in the solution of mathematical problems according to the Method, or the solution of other problems which though pertaining to other sciences, I was able to make almost similar to those of mathematics, by detaching them from all principles of other sciences which I found to be not sufficiently secure. You will see the result in many examples which are expounded in this volume. And hence, without living to all appearance in any way differently from those who, having no occupation beyond spending their lives in ease and innocence, study to separate pleasure from vice, and who, in order to enjoy their leisure without weariness, make use of all distractions that are innocent and good, I did not cease to prosecute my design, and to profit perhaps even more in my study of truth than if I had done nothing but read books or associate with literary people.

These nine years thus passed away before I had taken any definite part in regard to the difficulties as to which the learned are in the habit of disputing, or had commenced to seek the foundation of any philosophy more certain than the vulgar. And the example of many excellent men who had tried to do the same before me, but, as it appears to me, without success, made me imagine it to be so hard that possibly I should not have dared to undertake the task, had I not discovered that someone hand spread abroad the report that I had already reached its conclusion. I cannot tell on what they based this opinion; if my conversation has contributed anything to it, this must have arisen from my confessing my ignorance more ingenuously than those who have studied a little usually do. And perhaps it was also due to my having shown forth my reasons for doubting many things which were held by others to be certain, rather than from having boasted of any special philosophic system. But being at heart honest enough not to desire to be esteemed as different from what I am, I thought that I must try by every means in my power to render myself worthy of the reputation which I had gained. And it is just eight years ago that this desire made me resolve to remove myself from all places where any acquaintances were possible, and to retire to a country such as this, where the long-continued war has caused such order to be established that the armies which are maintained seem only to be of use in allowing the inhabitants to enjoy the fruits of peace with so much the more security; and where, in the crowded throng of a great and very active nation, which is more concerned with its own affairs than curious about those of others, without missing any of the conveniences of the most populous towns, I can live as solitary and retired as in deserts the most remote.

PART IV

I do not know that I ought to tell you of the first meditations there made by me, for they are so metaphysical and so unusual that they may perhaps not be acceptable to everyone. And yet at the same time, in order that one may judge whether the foundations which I have found are sufficiently secure, I find myself constrained in some measure to refer to them. For a long time I had remarked that it is sometimes requisite in common life to follow opinions which one knows to be most uncertain, exactly as though they were indisputable, as has been said above. But because in this case I wished to give myself entirely to the search after truth, I thought that it was necessary for me to take an apparently opposite course, and to reject as absolutely false everything as to which I could imagine the least ground of doubt, in order to see if afterwards there remained anything in my belief that was entirely certain. This, because our senses sometimes deceive us, I wished to suppose that nothing is just as they cause us to imagine it to be; and because there are men who deceive themselves in their reasoning and fall into paralogisms, even concerning the simplest matters of geometry, and judging that I was as subject to error as was any other, I rejected as false all the reasons formerly accepted by me as demonstrations. And since all the same thoughts and conceptions which we have while awake may also come to us in sleep, without any of them being at that time true, I resolved to assume that everything that ever entered into my mind was no truer than the illusions of my dreams. But immediately afterwards I noticed that whilst I thus wished to think all things false, it was absolutely essential that the 'I' who thought this should be somewhat, and remarking that this truth "*I think therefore I am*" was so certain and so assured that all the most extravagant suppositions brought forward by the skeptics were incapable of shaking it, I came to the conclusion that I could receive it without scruple as the first principle of the Philosophy for which I was seeking.

And then, examining attentively that which I was, I saw that I could conceive that I had no body, and that there was no world nor place where I might be; but yet that I could not for all that conceive that I was not. On the contrary, I saw from the very fact that I thought of doubting the truth of other things, it very evidently and certainly followed that I was; on the other hand if I had only ceased from thinking, even if all the rest of what I had ever imagined had really existed, I should have no reason for thinking that I had existed. From that I knew that I was a substance the whole essence or nature of which is to think, and that for its existence there is no need of any place, nor does it depend on any material thing; so that this 'me', that is to say, the soul by which I am what I am, is entirely distinct from body, and is even more easy to know than is the latter; and even if body were not, the soul would not cease to be what it is.

After this I considered generally what in a proposition is requisite in order to be true and certain; for since I had just discovered one which I knew to be such, I thought that I ought also to know in what this certainty consisted. And having remarked that there was nothing at all in the statement "*I think, therefore I am*" which assures me of having thereby made a true assertion, excepting that I see very clearly that to think it is necessary to be, I came to the conclusion that I might assume, as a general rule, that the things which we conceive very clearly and distinctly are all true - remembering, however, that there is some difficulty in ascertaining which are those that we distinctly conceive.

Following upon this, and reflecting on the fact that I doubted, and that consequently my existence was not quite perfect (for I saw clearly that it was a greater perfection to know than to doubt), I resolved to inquire whence I had learnt to think of anything more perfect than I myself was; and I recognized very clearly that this conception must proceed from some nature which was really more perfect. As to the thoughts which I had of many other things outside of me, like the heavens, the earth, light, heat, and a thousand others, I had not so much difficulty in knowing whence they came, because, remarking nothing in them which seemed to render them superior to me, I could believe that, if they were true, they were dependencies upon my nature, in so far as it possessed some perfection; and if they were not true, that I held them from naught, that is to say, that they were in me because I had something lacking in my nature. But this could not apply to the idea of a Being more perfect than my own, for to hold it from naught would be manifestly impossible; and because it is no less contradictory to say of the more perfect that it is what results from and depends on the less perfect, than to say that there is something which proceeds from nothing, it was equally impossible that I should hold it from myself. In this way it could but follow that it had been placed in me by a Nature which was really more perfect than mine could be, and which even had within it all the perfections of which I could form any idea - that is to say, to put it in a word, which was God. To which I added that since I knew some perfections which I did not possess, I was not the only being in existence (I shall here use freely, if you will allow, the terms of the School); but that there was necessarily some other more perfect Being on which I depended, or from which I acquired all that I had. For if I had existed alone and independent of any others, so that I should have had from myself all that perfection of being in which I participated to however small an extent, I should have been able for the same reason to have had all the remainder which I knew that I lacked; and thus I myself should have been infinite eternal, immutable, omniscient, all-powerful, and, finally, I should have all the perfections which I could discern in God. For, in pursuance of the reasoning which I have just carried on, in order to know the nature of God as far as my nature is capable of knowing it, I had only to consider in reference to all these things of which I found some idea in myself, whether it was a perfection to possess them or not. And I was assured that none of those which indicated some imperfection were in Him, but that all else was present; and I saw that doubt, inconstancy, sadness, and such things, could not be in Him considering that I myself should have been glad to be without them. In addition to this, I had ideas of many things which are sensible and corporeal, for, although I might suppose that I was dreaming, and that all that I saw or imagined was false, I could not at the same time deny that the ideas were really in my thoughts. But because I had already recognized very clearly in myself that the nature of the intelligence is distinct from that of the body, and observing that all composition gives evidence of dependency, and that dependency is manifestly an imperfection, I came to the conclusion that it could not be a perfection in God to be composed of these two natures, and that consequently He was not so composed. I judged, however, that if there were any bodies in the world, or even any intelligences or other natures which were not wholly perfect, their existence must depend on His power in I such a way that they could not subsist without Him for a single moment.

After that I desired to seek for other truths, and having put before myself the object of the geometricians, which I conceived to be a continuous body, or a space indefinitely extended in length, breadth, height or depth, which was divisible into various parts, and which might have various figures and sizes, and might be moved or transposed in all sorts of ways (for all this the geometricians suppose to be in the object of their contemplation) I went through some of their simplest demonstrations, and having noticed that this great certainty which everyone attributes to these demonstrations is founded solely on the fact that they are conceived of with clearness, in accordance with the rule which I have just laid down, I also noticed that there was nothing at all in them to assure me of the existence of their object. For, to take an example, I saw very well that if we suppose a triangle to be given, the three angles must certainly be equal to two right angles, but for all that I saw no reason to be assured that there was any such triangle in existence, while on the contrary, on reverting to the examination of the idea which I had of a Perfect Being, I found that in this case existence was implied in it in the same manner in which the equality of its three angles to two right angles is implied in the idea of a triangle; or in the idea of a sphere, that all the points on its surface are equidistant from its centre, or even more evidently still. Consequently it is at least as certain that God, who is a being so perfect, is, or exists, as any demonstration of geometry could possibly be.

What causes many, however, to persuade themselves that there is difficulty in knowing this truth, and even in knowing the nature of their soul, is the fact that they never raise their minds above the things of sense, or that they are so accustomed to consider nothing excepting by imagining it, which is a mode of thought specially adapted to material objects, that all that is not capable of being imagined appears to them not to be intelligible at all. This is manifest enough from the fact that even the philosophers in the Schools hold it as a maxim that there is nothing in the understanding which has not first of all been in the senses, in which there is certainly no doubt that the ideas of God and of the soul have never been. And it seems to me that those who desire to make use of their imagination in order to understand these ideas, act in the same way as if, to hear sounds or smell odors, they should wish to make use of their eyes: excepting that there is indeed this difference, that the sense of sight does not give us less assurance of the truth of its objects, than do those of scent or of hearing, while neither our imagination not our senses can ever assure us of anything, if our understanding does not intervene.

If there are finally any persons who are not sufficiently persuaded of the existence of God and of their soul by the reasons which I have brought forward, I wish that they should know that all other things of which they perhaps think themselves more assured (such as possessing a body, and that there are stars and an earth and so on) are less certain. For, although we have a moral assurance of these things which is such that it seems that it would be extravagant in us to doubt them, at the same time no one, unless he is devoid of reason, can deny, when a metaphysical certainty is in question, that there is sufficient cause for our not having complete assurance, by observing the fact that when asleep we may similarly imagine that we have another body, and that we see other stars and another earth, without there being anything, of the kind.

For how do we know that the thoughts that come in dreams are more false than those that we have when we are awake, seeing that often enough the former are not less lively and vivid than the latter And though the wisest minds may study the matter as much as they will, I do not believe that they will be able to give any sufficient reason for removing this doubt, unless they presuppose the existence of God. For to begin with, that which I have just taken as a rule, that is to say, that all the things that we very clearly and very distinctly conceive of are true, is certain only because God is or exists, and that He is a. Perfect Being, and that all that is in us issues from Him. From this it follows that our ideas or notions, which to the extent of their being clear or distinct are ideas of real things issuing from God, cannot but to that extent be true. So that though we often enough have ideas which have an element of falsity, this can only be the case in regard to those which have in them somewhat that is confused or obscure, because in so far as they have this character they participate in negation - that is, they exist in us as confused only because we are not quite perfect. And it is evident that there is no less repugnance in the idea that error or imperfection, inasmuch as it is imperfection, proceeds from God, than there is in the idea of truth or perfection proceeding from naught. But if we did not know that all that is in us of reality and truth proceeds from a perfect and infinite Being, however clear and distinct were our ideas, we should not have any reason to assure ourselves that they had the perfection of being true.

But after the knowledge of God and of the soul has thus rendered us certain of this rule, it is very easy to understand that the dreams which we imagine in our sleep should not make us in any way doubt the truth of the thoughts which we have when awake. For even if in sleep we had some very distinct idea such as a geometrician might have who discovered some new demonstration the fact of being asleep would not annihilate against its truth. And as to the most ordinary error in our dreams, which consists in their representing to us various objects in the same way as do our external senses, it does not matter that this should give us occasion to suspect the truth of such ideas, because we may be likewise often enough deceived in them without our sleeping at all, just as when those who have the jaundice see everything as yellow, or when stars or other bodies which are very remote appear much smaller than they really are, or, finally, whether we are awake or asleep, we should never allow ourselves to be persuaded excepting by the evidence of our

Reason. And it must be remarked that I speak of our Reason and not of our imagination nor of our senses; just as though we see the sun very clearly, we should not for that reason judge that it is of the size of which it appears to be; likewise we could quite well distinctly imagine the head of a lion on the body of a goat, without necessarily concluding that a chimera exists. Reason does not insist that whatever we see or imagine thus is a truth, but it tells us clearly that all our ideas or notions must have some foundation of truth. Otherwise it could not be possible that God, who is all perfection and truth, should have placed them within us. And because our reasons are never so evident nor so complete during sleep as during wakefulness, although sometimes our imaginations are then just as lively and acute, or even more so, Reason tells us that since our thoughts cannot possibly be all true, because we are not altogether perfect, that which they have of truth must infallibly be met with in our waking experience rather than in that of our dreams.

PART V

I should be very glad to proceed to show forth the complete chain of truths which I have deduced from these first, but because to do this it would have been necessary now to speak of many matters of dispute among the learned, with whom I have no desire to embroil myself, I think that it will be better to abstain. I shall only state generally what these truths are, so that it may be left to the decision of those best able to judge whether it would be of use for the public to be more particularly informed of them or not. I always remained firm in the resolution which I had made, nor to assume any other principle than that of which I have just made use, in order to demonstrate the existence of God and of the Soul, and to accept nothing as true which did not appear to be more clear and more certain than the demonstrations of the geometricians had formerly seemed. And nevertheless I venture to say that not only have I found the means of satisfying myself in a short time as to the more important of those difficulties usually dealt with in philosophy, but I have also observed certain laws which God has so established in Nature, and of which He has imprinted such ideas on our minds, that, after having reflected sufficiently upon the matter, we cannot doubt their being accurately observed in all that exists or is alone in the world. Further, in considering the sequence of these laws, it seems to me that I have discovered many things more useful and more important than all that I had formerly learned or even hoped to learn.

But because I tried to explain the most important of these in a Treatise which certain considerations prevented me from publishing I cannot do better, in making them known, than here summarize briefly what that Treatise contains. I had planned to comprise in it all that I believed myself to know regarding the nature of material objects, before I set myself to write...

Chapter 4

GALILEO'S CONSIDERATIONS ON THE COPERNICAN OPINION

Galileo Galeli

I

In order to remove (as much as the blessed God allows me) the occasion to deviate from the most correct judgment about the resolution of the pending controversy, I shall try to do away with two ideas. These are notions which I believe some are attempting to impress on the minds of those persons who are charged with the deliberations, and, if I am not mistaken, they are concepts far from the truth.

The first is that no one has any reason to fear that the outcome might be scandalous; for the earth's stability and sun's motion are so well demonstrated in philosophy that we can be sure and indubitably certain about them; on the other hand, the contrary position is such an immense paradox and obvious foolishness so that no one can doubt in any way that it cannot be demonstrated now or ever, or indeed that it can never find a place in the mind of sensible persons. The other idea which they try to spread is the following: although that contrary assumption has been used by Copernicus and other astronomers, they did this in a suppositional manner, and insofar as it can account more conveniently for the appearances of celestial motions and facilitate astronomical calculations and computations, and it is not the case that the same persons who assumed it believed it to be true de facto and in nature; so the conclusion is that one can safely proceed to condemn it. However, if I am not mistaken, these ideas are fallacious and far from the truth, as I can show with the following considerations. These will only be general and suitable to be understood without much effort and labor even by someone who is not well versed in the natural and astronomical sciences. For, if there were the opportunity to treat these points with those who are very experienced in these studies, or at least who have the time to do the work required by the difficulty of the subject, then I should propose nothing but the reading of Copernicus's own book; from it and from the strength of his demonstrations one could clearly sec how true or false are the two ideas we are discussing.

That it is not to be disparaged as ridiculous is, therefore, clearly shown by the quality of the men, both ancient and modern, who have held and do hold it. No one can regard it as ridiculous unless he considers ridiculous and foolish Pythagoras with all his school, Philolaus (teacher of Plato), Plato himself (as Aristotle testifies in his book *On the Heavens*), Heraclites of Pontus, Ecphantus, Aristarchus of Samos, Hicetas and Seleucus the mathematician. Seneca himself not only does not ridicule it, hut he makes fun of those who do, writing in his book *On Comets*: "It is also important to study these questions in order to learn whether the universe goes around the motionless earth, or the earth rotates but the universe does not. Some have said that we are naturally unaware of motion that sunrise and sunset are not due to the motion of the heavens, but that it is we ourselves who rise and set. The matter deserves consideration, so that we may know the conditions of our existence, whether we stand still or move very fast, whether God drives everything around us or drives us." Regarding the moderns, Nicolas Copernicus first accepted it and amply confirmed it in his whole book. Then there were others: William Gilbert, a distinguished physician and philosopher, who treats it at length and confirms it in his book *On the Loadstone*; Johannes Kepler, a living illustrious philosopher and mathematician in the service of the former and the current Emperor, follows the same opinion; Origanus (David Tost) at the beginning of his Ephemerides supports the earth's motion with a very long discussion; and there is no lack of other authors who have published their reasons on the matter. Furthermore, though they have not published anything, I could name very many followers of this doctrine living in Rome, Florence, Venice, Padua, Naples, Pisa, Parma, and other places. This doctrine is not, therefore, ridiculous, having been accepted by great men; and, though their number is small compared to the followers of the common position, this is an indication of its being difficult to understand, rather than of its absurdity. Moreover, that it is grounded on very powerful and effective reasons may be shown from the fact that all its followers were previously of the contrary opinion, and indeed that for a long time they laughed at it and considered it foolish. Copernicus and I, and all others who are alive, are witnesses to this. Now, who will not believe that an opinion which is considered silly and indeed foolish, which has hardly one out of a thousand philosophers following it, and which is disapproved by the Prince of the prevailing philosophy, can become acceptable through anything but very firm demonstrations, very clear experiences, and very subtle observations? Certainly no one will be dissuaded of an opinion imbibed with mother's milk from his earliest training, accepted by almost the whole world and supported by the authority of very serious writers, unless the contrary reasons are more than effective. If we reflect carefully, we find that there is more value in the authority of a single person who follows the Copernican opinion than in that of one hundred others who hold the contrary, since those who are persuaded of the truth of the Copernican system were in the beginning all very opposed. So I argue as follows.

Those who are to be persuaded are capable of understanding the reasons of Copernicus and others, who follow him, or they are not; moreover, either these reasons are true and demonstrative, or they are fallacious. If those who are to be persuaded are incapable, then they will never be persuaded by the true or by the false reasons; those who are capable of understanding the strength of the demonstrations will likewise never be persuaded if these demonstrations are fallacious; so neither those who do nor those who do not understand will be persuaded by fallacious reasons. Therefore, given that absolutely no one can be dissuaded from the first idea by fallacious reasons, it follows as a necessary consequence that, if anyone is persuaded of the contrary of what he previously believed, the reasons arc persuasive and true. But as a matter of fact there are many who are already persuaded by Copernican reasons. Therefore, it is true both that these reasons are effective and that the opinion does not deserve the label of ridiculous but the label of worthy of being very carefully considered and pondered.

Furthermore, how futile it is to argue for the plausibility of this or that opinion simply from the large number of followers may be easily inferred from this: no one follows this opinion who did not previously believe the contrary; but instead you will not find even a single person who, after holding this opinion, will pass to the other one, regardless of any discussion he hears; consequently, one may judge, even if he does not understand the reasons for one side or for the other, that probably the demonstrations for the earth's motion are stronger than those for the other side. But I shall say more, namely that if the probability of the two positions were something to be won by ballot,

I would be willing to concede defeat when the opposite side had one more vote than I out of one hundred; not only that, but I would be willing to agree that every individual vote of the opponents was worth ten of mine, as long as the decision was made by persons who had perfectly heard, intimately penetrated, and subtly examined all the reasons and evidence of the two sides; indeed it is reasonable to expect that such would be those who cast the votes. Hence this opinion is not ridiculous and contemptible, but somewhat shaky is the position of whoever wanted to capitalize on the common opinion of the many who have not accurately studied these authors. What then should we say of the noises and the idle chatter of someone who has not understood even the first and simplest principles of these doctrines, and who is not qualified to understand them ever? What importance should we give him?

Consider now those who persist in wanting to say that as an astronomer Copernicus considered the earth's motion and the sun's stability only a hypothesis which is more adequate to save celestial appearances and to calculate the motions of planets, but that he did not believe it to be true in reality and in nature. With all due respect, these people show that they have been too prone to believe the word of someone who speaks more out of whim than out of experience with Copernicus's book or understanding the nature of this business. For this reason they talk about it in a way that is not altogether right.

First, limiting ourselves to general considerations, let us see his preface to Pope Paul III, to whom he dedicates the work. We shall find, to begin with, as if to comply with what they call the astronomer's task, that he had done and completed the work in accordance with the hypothesis of the prevailing philosophy and of Ptolemy himself, so that there was in it nothing lacking. But then, taking off the clothes of a pure astronomer and putting on those of a contemplator of nature, he undertook to examine whether this astronomical assumption already introduced, which was completely satisfactory regarding the calculations and the appearances of the motions of all planets, could also truly happen in the world and in nature. He found that in no way could such an arrangement of parts exist: although each by itself was well proportioned, when they were put together the result was a very monstrous chimera.

And so he began to investigate what the system of the world could really be in nature, no longer for the sole convenience of the pure astronomer, whose calculations he had complied with, but in order to come to an understanding of such a noble physical problem; he was confident that, if one had been able to account for mere appearances by means of hypotheses which are not true, this could be done much better by means of the true and physical constitution of the world. Having at his disposal a very large number of physically true and real observations of the motions of the stars (and without this knowledge it is wholly impossible to solve the problem), he worked tirelessly in search of such a constitution. Encouraged by the authority of so many great men, he examined the motion of the earth and the stability of the sun. Without their encouragement and authority, by himself either he would not have conceived the idea, or he would have considered it a very great absurdity and paradox, as he confesses to have considered it at first. But then, through long sense observations, favorable results, and very firm demonstrations, he found it so consonant with the harmony of the world that he became completely certain of its truth. Hence this position is not introduced to satisfy the pure astronomer, but to satisfy the necessity of nature.

Furthermore, Copernicus knew and wrote in the same place that publishing this opinion would have made him look insane to the numberless followers of current philosophy, and especially to each and every layman. Nevertheless, urged by the requests of the Cardinal of Capua and the Bishop of Klum, he published it. Now, would he not have been truly mad if, considering this opinion physically false, he had published that he believed it to be true, with the certain consequence that he would be regarded as a fool by the whole world? And why would he not have declared that he was using it only as an astronomer, but that he denied it as a philosopher, thus escaping the universal label of foolishness, to the advantage of his common sense?

Moreover, Copernicus states in great detail the grounds and reasons why the ancients believed the earth to be motionless, and then, examining the value of each in turn, he shows them to be ineffective. Now, who ever saw a sensible author engaged in confuting the demonstrations that confirm a proposition he considers true and real? And what kind of judgment would it be to criticize and to condemn a conclusion while in reality he wanted the reader to believe that he accepted it? This sort of incoherence cannot be attributed to such a man. Furthermore, note carefully that, since we are dealing with the motion or stability of the earth or of the sun, we are in a dilemma of contradictory propositions (one of which has to be true), and we cannot in any way resort to saying that perhaps it is neither this way nor that way. Now, if the earth's stability and sun's motion are de facto physically true and the contrary position is absurd, how can one reasonably say that the false view agrees better than the true one with the phenomena clearly visible and sensed in the movements and arrangement of the stars? Who does not know that there is a most agreeable harmony among all truths of nature, and a most sharp dissonance between false positions and true effects? Will it happen, then, that the earth's motion and sun's stability agree in every way with the arrangement of all other bodies in the universe and with all the phenomena, a thousand of them, which we and our predecessors have observed in great detail, and that this position is false? And can the earth's stability and sun's motion be considered true and not agree in any way with the other truths? If one could say that neither this nor that position is true, it might happen that one would be more convenient than the other in accounting for the appearances. But, given two positions, one of which must be true and the other false, to say that the false one agrees better with the effects of nature is really something that surprises my imagination. I add: if Copernicus confesses to having fully satisfied astronomers by means of the hypothesis commonly accepted as true, how can one say that by means of the false and foolish one he could or would want to satisfy again the same astronomers?

However, I now go on to consider the nature of the business from an internal viewpoint, and to show with how much care one must discuss it.

Astronomers have so far made two sorts of suppositions: some are primary and pertain to the absolute truth of nature; others are secondary and are imagined in order to account for the appearances of stellar motions, which appearances seem not to agree with the primary and true assumptions. For example, before trying to account for the appearances, acting not as a pure astronomer but as a pure philosopher, Ptolemy supposes, indeed he takes from philosophers, that celestial movements are all circular and regular, namely uniform; that heaven has a spherical shape; that the earth is at the center of the celestial sphere, is spherical, motionless, etc. Turning then to the inequalities we see in planetary movements and distances, which seem to clash with the primary physical suppositions already established, he goes on to another sort of supposition; these aim to identify the reasons why, without changing the primary ones, there is such a clear and sensible inequality in the movements of planets and in their approaching and their moving away from the earth. To do this he introduces some motions those are still circular, but around centers other than the earths, tracing eccentric and epicyclical circles. This secondary supposition is the one of which it could be said that the astronomer supposes it to facilitate his computations, without committing himself to maintaining that it is true in reality and in nature.

Let us now see in what kind of hypothesis Copernicus places the earth's motion and sun's stability. There is no doubt whatever, if we reflect carefully, that he places them among the primary and necessary suppositions about nature. For, as I have already stated, it seems that he had already given satisfaction to astronomers by the other road, and that he takes this one only to try to solve the greatest problem of nature. In fact, to say that he makes this supposition to facilitate astronomical calculations is so false that instead we can see him, when he comes to these calculations, leaving this supposition and returning to the old one, the latter being more readily and easily understood and still very quick even in computations. This may be seen as follows. Intrinsically, particular calculations can be made by taking one position as well as the other, that is, by making the earth or the heavens rotate; nevertheless, many geometers and astronomers in many books have already demonstrated the properties of orthogonal and oblique displacements of parts of the zodiac in relation to the equator, the declinations of the parts of the ecliptic, the variety of angles between it and both meridians and oblique horizons, and a thousand other specific details necessary to complete astronomical science. This ensures that, when he comes to examining these details of the primary motions, Copernicus himself examines them in the old manner, namely as occurring along circles traced in the heavens and around the motionless earth, even though stillness and stability should belong to the highest heaven, called the Prime Mobile, and motion to the earth. Thus in the introduction to Book Two he concludes: "People should not be surprised if we still use the ordinary terms for the rising and setting of the Sun and stars and similar occurrences, but should recognize that we are speaking in customary language, which is acceptable to everyone, yet always bearing in mind that 'For us who ride the Earth, the Sun and Moon are passing; patterns of stars return, and then again recede'." We should therefore understand clearly that Copernicus takes the earth's motion and sun's stability for no other reason and in no other way than to establish it, in the manner of the natural philosopher, as a hypothesis of the primary sort; on the contrary, when he comes to astronomical computations, he goes back to the old hypothesis, which takes the circles of the basic motions with their details to be located in the highest heaven around the motionless earth, being easier for everyone to understand on account of ingrained habit. But what am I saving? Such is the strength of truth and the weakness of falsehood, that those who speak this way reveal themselves not completely capable of understanding these subjects and not well versed in them; this happens when they let themselves be persuaded that the secondary kind of hypothesis is considered chimerical and fictional by Ptolemy and by other serious astronomers, and that they really regard them as physically false and introduced only for the sake of astronomical computations. The only support they give for this very fanciful opinion is a passage in Ptolemy where, unable to observe more than one simple anomaly in the sun, he wrote that to account for it one could take the hypothesis of a simple eccentric as well as that of an epicycle on a concentric, and he added he preferred the first for being simpler than the second; from these words some very superficially argue that Ptolemy did not consider necessary, but rather wholly fictional, both this and that supposition, since he said they are both equally convenient, while one and only one can be attributed to the sun's behavior. But what kind of superficiality is this? Who can do both of the following? First, to suppose as true the primary suppositions that planetary motions are circular and regular, and to admit (as the senses themselves necessarily force us) that in running through the zodiac all planets are now slow and now fast, indeed that most of them can be not only slow but also stationary and retrograde, and that we see them now very large and very near the earth and now very small and very far; and then, having understood these former points, to deny that eccentrics and epicycles can really exist in nature? This is wholly excusable for men who are not specialists in these sciences, but for others who would claim to be experts in them it would be an indication that they do not even understand the meaning of the terms eccentric and epicycle. One might just as well first admit that there are three letters, the first of which is G, the second O, and the third D, and then at the end deny that their combination yields GOD and claim that the result is SHADOW. But if rational arguments were not sufficient to make one understand the necessity of having to place eccentrics and epicycles really in nature, at least the senses themselves would have to persuade him: for we see the four Medicean planets' trace four small circles around Jupiter which are very far from enclosing the earth, in short, four epicycles; Venus, which is seen now full of light and now very thinly crescent, provides conclusive evidence that its revolution is around the sun and not around the earth, and consequently that its orbit is an epicycle; and the same may be argued for the case of Mercury. Moreover, the three outer planets are very near the earth when they are in opposition to the sun, and very far when in conjunction; for example, Mars at its closest appears to the senses more than fifty times larger than at its farthest, so that some have occasionally feared that it had gotten lost or had vanished, being really invisible because of its great distance; now, what else can one conclude but that their revolution is made in eccentric circles, or in epicycles, or in a combination of the two, if we take the second anomaly into consideration? So, to deny eccentrics and epicycles in the motions of planets is like denying the light of the sun, or else it is to contradict oneself. Let us apply what I am saying more directly to our purpose: some say that modern astronomers introduce the earth's motion and sun's stability suppositional in order to account for the phenomena and to facilitate calculations, just as epicycles and eccentrics are assumed in the same manner, though the same astronomers consider them physically chimerical and repugnant; I answer that I shall gladly agree with all this talk, as long as they limit themselves to staying within their own conceptions, namely that the earth's motion and sun's stability are as false or true in nature as epicycles and eccentrics. Let them, then, make every effort to do away with the true and real existence of these circles, for if they succeed in demonstrating their non-existence in nature, I shall immediately surrender and admit the earth's motion to be a great absurdity. But if, on the contrary, they are forced to accept them, let them also accept the earth's motion, and let them admit to have been convinced by their own contradictions.

I could present many other things for this same purpose. However, since I think that whoever is not persuaded by what I have said would not be persuaded by many more reasons either, I want these to suffice. I shall only add something about what could have been the motive why some have concluded with any plausibility that Copernicus himself did not really believe his own hypothesis.

There is on the reverse side of the title page of Copernicus's book a certain preface to the reader, which is not by the author since it mentions him in the third person and is without signature. It clearly states that no one should believe in the least that Copernicus regarded his position as true, but only that he feigned and introduced it for the calculation of celestial motions; it ends its discussion by concluding that to hold it as true and real would be foolish. This conclusion is so explicit that whoever reads no further, and believes it to have been placed at least with the author's consent, deserves to be somewhat excused for his error. But what weight to give to the opinion of those who would judge a book without reading anything but a brief preface by the printer or publisher, I let each one decide for himself. I say that this preface can only have originated from the publisher to facilitate the sale of a book which common people would have regarded as a fanciful chimera if a similar preface had not been added; for most of the time buyers are in the habit of reading such prefaces before buying the work. Not only was this preface not written by the author, but it was included without his consent, and also without his knowledge; this is shown by the errors it contains, which the author would never have committed.

This preface says no one can consider it verisimilar, unless he is completely ignorant of geometry and optics, that Venus has such a large epicycle enabling it now to precede and now to follow the sun by 40 degrees or more; for it would have to happen that when it is highest its diameter should appear only one-fourth of what it appears when it is lowest, and that in the latter location its body should be seen as sixteen times bigger than in the former; but these things, he says, are repugnant to the observations made throughout the centuries. In these assertions we see, first, that the writer does not know that Venus departs on one side and on the other of the sun by about 48 degrees and not 40 as he says. Moreover, he asserts that its diameter should appear four times, and its body sixteen times, larger in one position than in the other. Here, first, due to a geometrical oversight he does not understand that when one globe has a diameter four times larger than another, its body is sixty-four times bigger, and not sixteen, as he stated. Hence, if he considered such an epicycle absurd and wanted to declare it to be physically impossible, if he had understood this subject, he could have made the absurdity much greater; for, according to the position he wants to refute (well known to astronomers), Venus digresses from the sun almost 48 degrees, and when farthest from the earth its distance must be six times greater than when closest, and consequently its apparent diameter in the latter position is more than six times larger than in the former (not four times), and its body more than two hundred and sixteen times greater (and not just sixteen). These errors are so gross that it is impossible to believe they were committed by Copernicus, or by anyone else but the most unqualified persons. Moreover, why label such a large epicycle most absurd, so that because of such an absurdity we would conclude that Copernicus did not regard his assumptions as true, and that neither should others so regard them? He should have remembered that in chapter 10 of the first hook Copernicus is speaking ad hominem and is attacking other astronomers who allege that it is a great absurdity to give Venus such an epicycle, which is so large as to exceed the whole lunar orbit by more than two hundred times, and which does not contain anything inside; he then removes the absurdity when he shows that inside Venus's orbit is contained the orbit of Mercury and, placed at the centre, the body of the sun itself. What frivolity is this, then, to want to show a position mistaken and false on account of a difficulty which that position not only does not introduce in nature, but completely removes? Similarly it removes the immense epicycles which out of necessity other astronomers assumed in the other system. This only touches the writer of Copernicus's preface; so we may argue that, if he had included something else professionally relevant, he would have committed other errors.

But finally, to remove any shadow of a doubt, if the failure to observe such great variations in the apparent sizes of the body of Venus should cast doubt on its circular revolution around the sun (from the viewpoint of the Copernican system), then let us make careful observations with a suitable instrument, namely with a good telescope, and we shall find all effects and experiences exactly agreeing; that is, we shall see Venus crescent when it is nearest to the earth, and with a diameter six times larger than when it is at its maximum distance, namely above the sun, where it is seen round and very small. I have discussed elsewhere the reasons for not detecting these variations with our simple eyesight, but just as from this failure we could reasonably deny that supposition, so now, from seeing the very exact correspondence in this and every other detail, we should abandon any doubt and consider the supposition true and real. As for the rest of this admirable system, whoever desires to ascertain the opinion of Copernicus himself should not read the fanciful preface of the printer, but the whole work of the author himself; without a doubt he will grasp first-hand that Copernicus held as very true the stability of the sun and the motion of the earth.

II

The motion of the earth and stability of the sun could never be against Faith or Holy Scripture, if this proposition were correctly proved to be physically true by philosophers, astronomers, and mathematicians, with the help of sense experiences, accurate observations, and necessary demonstrations. However, in this case, if some passages of Scripture were to sound contrary, we would have to say that this is due to the weakness of our mind, which is unable to grasp the true meaning of Scripture in this particular case. This is the common doctrine, and it is entirely right, since one truth cannot contradict another truth. On the other hand, whoever wants to condemn it judicially must first demonstrate it to be physically false by collecting the reasons against it.

Now, one wants to know where to begin in order to ascertain its falsity, that is, whether from the authority of Scripture or from the refutation of the demonstrations and observations of philosophers and astronomers. I answer that one must start from the place which is safest and least likely to bring about a scandal; this means beginning with physical and mathematical arguments. For if the reasons proving the earth's motion arc found fallacious, and the contrary ones conclusive, then we have already become certain of the falsity of this proposition and the truth of the opposite, which we now say corresponds to the meaning of Scripture; so one would be free to condemn the false proposition and there would be no danger. But if those reasons arc found true and necessary, this will not bring any harm to the authority of Scripture; instead we shall have been cautioned that due to our ignorance we had not grasped the true sense of Scripture, and that we can learn this meaning with the help of the newly acquired physical truth. Therefore, beginning with the arguments is safe in any case. On the other hand, if we were to fix only on what seemed to us the true and certain meaning of Scripture, and we were to go on to condemn such a proposition without examining the strength of the arguments, what a scandal would follow if sense experiences and reasons were to show the opposite? And who would have brought confusion to the Holy Church? Who had suggested the greatest consideration of the arguments, or those who had disparaged them? One can see, then, which road is safer.

Moreover, we admit that a physical proposition which has been proved true by physical and mathematical demonstrations can never contradict Scripture, but that in such a case it is the weakness of our mind which prevents us from grasping its true meaning. On the other hand, whoever wants to use the authority of the same passages of Scripture to confute and prove false the same proposition would commit the error called "begging the question?" For, the true meaning of Scripture being in doubt in the light of the arguments, one cannot take it as clear and certain in order to refute the same proposition; instead one must cripple the arguments and find the fallacies with the help of other reasons and experiences and more certain observations. When the factual and physical truth has been found in this manner, then, and not before, can one be assured of the true meaning of Scripture and safely use it. Thus the safe road is to begin with the arguments, confirming the true and refuting the fallacious ones.

If the earth de facto moves, we cannot change nature and arrange for it not to move. But we can rather easily remove the opposition of Scripture with the mere admission that we do not grasp its true meaning. Therefore, the way to be sure not to err is to begin with astronomical and physical investigations, and not with scriptural ones.

I am always told that, in interpreting the passages of Scripture relevant to this point, all Fathers agree to the meaning which is simplest and corresponds to the literal meaning; hence, presumably, it is improper to give them another meaning or to change the common interpretation, because this would amount to accusing the Fathers of carelessness or negligence. I answer by admitting that the Fathers indeed deserve reasonable and proper respect, hut I add that we have an excuse for them very readily: it is that on this subject they never interpreted Scripture differently from the literal meaning, because at their time the opinion of the earth's motion was totally buried and no one even talked about it, let alone wrote about it or maintained it. But there is no trace of negligence by the Fathers for not thinking about what was completely hidden.

That they did not think about it is clear from the fact that in their writings one cannot find even a word about this opinion. And if anyone were to say that they considered it, this would make its condemnation more dangerous; for after considering it, not only did they not condemn it, but they did not express any doubt about it.

Thus the defense of the Fathers is readily available and very easy. On the contrary, it would be very difficult or impossible to excuse or exonerate from a similar charge of carelessness the Popes, Councils, and Congregations of the Index of the last eighty years, if this doctrine were erroneous and deserving of condemnation; for they have let this opinion circulate in a book which was first written on orders from a Pope, and then printed on orders from a cardinal and a bishop, dedicated to another Pope, and, most important, received by the Holy Church, so that one cannot say that it had remained unknown. If, then, the inappropriateness of charging our highest authorities with negligence is to be taken into account, as it should, Let us make sure that in trying to escape one absurdity we do not fall into a greater one.

But assume now that someone regards it as inappropriate to abandon the unanimous interpretation of the Fathers, even in the case of physical propositions not discussed by them and whose opposite they did not even consider; I then ask what one should do if necessary demonstrations showed the facts of nature to be the opposite. Which of the two decrees should be changed? The one which stipulates that no proposition can be both true and erroneous, or the other one which obliges us to regard as articles of faith physical propositions supported by the unanimous interpretation of the Fathers?

It seems to me, if I am not mistaken, that it would be safer to modify this second decree than to be forced to hold as an article of faith a physical proposition which had been demonstrated with conclusive reasons to be factually false in nature. It also seems to me that one could say that the unanimous interpretation of the Fathers should have absolute authority in the case of propositions which they aired, and for which no contrary demonstrations exist and it is certain that none could ever exist.

I do not bring in the fact that it is very clear that the Council requires only that one agree with the unanimous interpretation of the Fathers "in matters of faith and morals, etc."

III

1. Copernicus uses eccentrics and epicycles, but these were not the reason for rejecting the Ptolemaic system, since they undoubtedly exist in the heavens; it was other difficulties.
2. In regard to philosophers, if they were true philosophers, namely lovers of truth, they should not get irritated, but, learning that they were wrong, they should thank whoever shows them the truth; and if their opinion were to stand up, they would have reason to take pride in it, rather than being irritated. Theologians should not get irritated because, if this opinion were found false then they could freely prohibit it, and if it were discovered true then they should rejoice that others have found the way to understand the true meaning of Scripture and have restrained them from perpetrating a serious scandal by condemning a true proposition. In regard to falsifying Scripture, this is not and will never be the intention of Catholic astronomers such as ourselves; rather our view is that Scripture corresponds very well to truths demonstrated about nature. Moreover, certain theologians who are not astronomers should be careful about falsifying Scripture by wanting to interpret it as opposed to propositions which may be true and demonstrable.
3. It might happen that we could have difficulties in interpreting Scripture, but this would occur because of our ignorance and not be because there really are or can be insuperable difficulties in reconciling Scripture with demonstrated truths.
4. The Council speaks "about matters of faith and morals, etc." So there is an answer to the claim that such a proposition is "an article of faith by reason of the speaker," though not "by reason of the topic," and that therefore it is among those covered by the Council. The answer is that everything in Scripture is "an article of faith by reason of the speaker," so that in this regard it should be included in the rule of the Council; but this clearly has not been done because in that case it would have said that "the interpretation of the Fathers is to be followed for every word of Scripture, etc.," and not "for matters of faith and morals"; having thus said "for matters of faith," we see that its intention was to mean "for matters of faith by reason of the topic." Then consider that it is much more a matter of faith to hold that Abraham had some children and that Tobias had a dog, because Scripture says it, than it would be to hold that the earth moves, even if

this were found in the same Scripture, and further that to deny the former is a heresy, but not to deny the latter. It seems to me that this depends on the following reason. There have always been in the world men who had two, four, six children, etc., or none, and similarly people who have dogs and who do not, so that it is equally credible that some have children or dogs and others do not; hence there appears to be no reason why in such propositions the Holy Spirit should speak differently from the truth, the negative and the affirmative sides being equally credible to all men. But it is not so with the motion of the earth and the stability of the sun; these propositions are very far removed from the understanding of the masses, for on these matters not relevant to their eternal life the Holy Spirit chose to conform its pronouncements with their abilities, even when facts are otherwise from the point of view of the thing in itself.

5. In regard to placing the sun in heaven and the earth outside it, as Scripture seems to affirm, etc., this truly seems to me to be a simple perception of ours and a manner of speaking only for our convenience. For, in reality all that is surrounded by heaven is in heaven, just as all that is surrounded by the city walls is in the city; indeed, if one were to express a preference, what is in the middle is more in heaven and in the city, being, as it were, at the heart of the city and of heaven. That difference exists because one takes the elemental region surrounding the earth as being very different from the celestial region. But such a difference will always exist regardless of where these elements are placed; and it will always be true that from the viewpoint of our convenience the earth is below us and heaven above, since all the inhabitants of the earth have heaven above our heads, which is our upwards, and the center of the earth under our feet, which is our downwards; so, in relation to us the center of the earth and the surface of heaven are the farthest places, that is, the endpoints of our up and down, which are diametrically opposite points.
6. Not to believe that there is a demonstration of the earth's mobility until it is shown is very prudent, nor do we ask that anyone believe such a thing without a demonstration. On the contrary, we only seek that, for the advantage of the Holy Church, one examine with the utmost severity what the followers of this doctrine know and can advance, and that nothing be granted them unless the strength of their arguments greatly exceeds that of the reasons for the opposite side.

Now, if they are not more than ninety percent right, they may be dismissed; but if all that is produced by philosophers and astronomers on the opposite side is shown to be mostly false and wholly inconsequential, then the other side should not he disparaged, nor deemed paradoxical, so as to think that it could never he clearly proved. It is proper to make such a generous offer since it is clear that those who hold the false side cannot have in their favor any valid reason or experiment, whereas it is necessary that all things agree and correspond with the true side.

7. It is true that it is not the same to show that one can save the appearances with the earth's motion and the sun's stability, and to demonstrate that these hypotheses are really true in nature.' But it is equally true, or even more so, that one cannot account for such appearances with the other commonly accepted system. The latter is undoubtedly false, while it is clear that the former, which can account for them, may be true. Nor can one or should one seek any greater truth in a position than that it corresponds with all particular appearances.
8. One is not asking that in case of doubt the interpretation of the Fathers should be abandoned, but only that an attempt be made to gain certainty regarding what is in doubt, and that therefore no one disparage what attracts and has attracted very great philosophers and astronomers. Then, after all necessary care has been taken, the decision may be made.
9. We believe that Solomon, Moses, and all other sacred writers knew perfectly the constitution of the world, as they also knew that God has no hands, no feet, and no experience of anger, forgetfulness, or regret; nor will we ever doubt this. But we say what the Holy Fathers and in particular St. Augustine say about these matters, namely that the Holy Spirit inspired them to write what they wrote for various reasons, etc.
10. The error of the apparent movement of the shore and stability of the ship is known by us after having many times observed the motion of boats from the shore, and many other times observed the shore from a boat; and so, if we could now stay on earth and now go to the sun or other star, perhaps we would acquire sensible and certain knowledge of which one of them moves. To be sure, if we looked only at these two bodies, it would always seem to us that the one we were on was standing still, just as looking only at the water and the boat always gives the appearance that the water is flowing and the boat is standing

still. Moreover, the two situations are very different: there is great disparity between a small boat, separable from its environment, and the immense shore, known by us through thousands of experiences to be motionless, that is, motionless in relation to the water and the boat; but the other comparison is between two bodies both of which are substantial and equally inclined toward motion and toward rest. Thus it would be more relevant to compare between themselves two boats, in which case it is absolutely certain that the one we were on would always appear to us as motionless, as long as we could not consider any other relationship but that which holds between these two ships.

There is, therefore, a very great need to correct the error about observing whether the earth or else the sun moves, for it is clear that to someone on the moon or any other planet it would always appear that it was standing still and the other stars were moving. But these and many other more plausible reasons of the followers of the common opinion are the ones that must be untied very openly, before one can pretend even to be heard, let alone approved; unfortunately we have not done a very detailed examination of what is produced against us. Moreover, neither Copernicus nor his followers will ever use this phenomenon of the shore and the boat to prove that the earth is in motion and the sun at rest. They only adduce it as an example that serves to show, not the truth of their position, but the absence of contradiction between the appearance of a stable earth and moving sun to our simple sense experience, and the reality of the contrary. For if this was one of Copernicus's demonstrations, or if his others did not argue more effectively, I really think that no one would agree with him.

Chapter 5

THE PHILOSOPHICAL IMPORTANCE OF MATHEMATICAL LOGIC

Bertrand Russell

In speaking of "Mathematical logic", I use this word in a very broad sense. By it I understand the works of Cantor on transfinite numbers as well as the logical work of Frege and Peano. Weierstrass and his successors have "arithmetised" mathematics; that is to say, they have reduced the whole of analysis to the study of integer numbers. The accomplishment of this reduction indicated the completion of a very important stage, at the end of which the spirit of dissection might well be allowed a short rest. However, the theory of integer numbers cannot be constituted in an autonomous manner, especially when we take into account the likeness in properties of the finite and infinite numbers. It was, then, necessary to go farther and reduce arithmetic, and above all the definition of numbers, to logic. By the name "mathematical logic", then, I will denote any logical theory whose object is the analysis and deduction of arithmetic and geometry by means of concepts which belong evidently to logic. It is this modern tendency that I intend to discuss here.

In an examination of the work done by mathematical logic, we may consider either the mathematical results, the method of mathematical reasoning as revealed by modern work, or the intrinsic nature of mathematical propositions according to the analysis which mathematical logic makes of them. It is impossible to distinguish exactly these three aspects of the subject, but there is enough of a distinction to serve the purpose of a framework for discussion. It might be thought that the inverse order would be the best; that we ought first to consider what a mathematical proposition is, then the method by which such propositions are demonstrated, and finally the results to which this method leads us. But the problem which we have to resolve, like every truly philosophical problem, is a problem of analysis; and in problems of analysis the best method is that which sets out from results and arrives at the premises. In mathematical logic it is the conclusions which have the greatest degree of certainty: the nearer we get to the ultimate premises the more uncertainty and difficulty do we find.

From the philosophical point of view, the most brilliant results of the new method are the exact theories which we have been able to form about infinity and continuity. We know that when we have to do with infinite collections, for example the collection of finite integer numbers, it is possible to establish a one-to-one correspondence between the whole collection and a part of itself. For example, there is such a correspondence between the finite integers and the even numbers, since the relation of a finite number to its double is one-to-one. Thus it is evident that the number of an infinite collection is equal to the number of a part of this collection. It was formerly believed that this was a contradiction; even Leibnitz, although he was a partisan of the actual infinite, denied infinite number because of this supposed contradiction. But to demonstrate that there is contradictions we must suppose that all numbers obey mathematical induction. To explain mathematical induction, let us call by the name "hereditary property" of a number a property which belongs to $n + 1$ whenever it belongs to n. Such is, for example, the property of being greater than 100. If a number is greater than 100, the next number after it is greater than 100. Let us call by the name "inductive property" of a number a hereditary property which is possessed by the number zero. Such a property must belong to 1, since it is hereditary and belongs to 0; in the same way, it must belong to 2, since it belongs to 1; and so on. Consequently the numbers of daily life possess every inductive property. Now, amongst the inductive properties of numbers is found the following. If any collection has the number n, no part of this collection can have the same number n. Consequently, if all numbers possess all inductive properties, there is a contradiction with the result that there are collections which have the same number as a part of themselves. This contradiction, however, ceases to subsist as soon as we admit that there are numbers which do not possess all inductive properties. And then it appears that there is no contradiction in infinite number. Cantor has even created a whole arithmetic of infinite numbers, and by means of this arithmetic he has completely resolved the former problems on the nature of the infinite which have disturbed philosophy since ancient times.

The problems of the *continuum* are closely connected with the problems of the infinite and their solution is affected by the same means. The paradoxes of Zeno the Eleatic and the difficulties in the analysis of space, of time, and of motion, are all completely explained by means of the modern theory of continuity. This is because a non-contradictory theory has been found, according to which the continuum is composed of infinity of distinct elements; and this formerly appeared impossible. The elements cannot all be reached by continual dichotomy; but it does not follow that these elements do not exist.

From this follows a complete revolution in the philosophy of space and time. The realist theories which were believed to be contradictory are so no longer, and the idealist theories have lost any excuse there might have been for their existence. The flux, which was believed to be incapable of analysis into indivisible elements, shows itself to be capable of mathematical analysis, and our reason shows itself to be capable of giving an explanation of the physical world and of the sensible world without supposing jumps where there is continuity, and also without giving up the analysis into separate and indivisible elements.

The mathematical theory of motion and other continuous changes uses, besides the theories of infinite number and of the nature of the continuum, two correlative notions, that of a *function* and that of a *variable*. The importance of these ideas may be shown by an example. We still find in books of philosophy a statement of the law of causality in the form: "When the same cause happens again, the same effect will also happen." But it might be very justly remarked that the same cause never happens again. What actually takes place is that there is a constant relation between causes of a certain kind and the effects which result from them. Wherever there is such a constant relation, the effect is a function of the cause. By means of the constant relation we sum up in a single formula an infinity of causes and effects, and we avoid the worn-out hypothesis of the *repetition* of the same cause. It is the idea of functionality, that is to say the idea of constant relation, which gives the secret of the power of mathematics to deal simultaneously with infinity of data.

To understand the part played by the idea of a function in mathematics, we must first of all understand the method of mathematical deduction. It will be admitted that mathematical demonstrations, even those which are performed by what is called mathematical induction, are always deductive. Now, in a deduction it almost always happens that the validity of the deduction does not depend on the subject spoken about, but only on the form of what is said about it. Take for example the classical argument: All men are mortal, Socrates is a man, and therefore Socrates is mortal. Here it is evident that what is said remains true if Plato or Aristotle or anybody else is substituted for Socrates. We can, then, say: If all men are mortal, and if *x* is a man, then *x* is mortal. This is a first generalization of the proposition from which we set out. But it is easy to go farther. In the deduction which has been stated, nothing depends on the fact that it is men and mortals which occupy our attention. If all the members of any class are members of a class *s*, and if *x* is a member of the class a, then *x* is a member of the class *s*. In this statement, we have the pure logical form which underlies all the deductions of the same form as that which proves that Socrates is mortal. To obtain a proposition of pure mathematics (or of mathematical logic, which is the same thing), we must submit a deduction of any kind to a process analogous to that which we have just performed, that is to say, when an argument remains valid if one of its terms is changed, this term must be replaced by a variable, i.e. by an indeterminate object. In this way we finally reach a proposition of pure logic, that is to say a proposition which does not contain any other constant than logical constants. The definition of the *logical constants* is not easy, but this much may be said: A *constant is logical* if the propositions in which it is found still contain it when we try to replace it by a variable. More exactly, we may perhaps characterize the logical constants in the following manner: If we take any deduction and replace its terms by variables, it will happen, after a certain number of stages, that the constants which still remain in the deduction belong to a certain group, and, if we try to push generalization still farther, there will always remain constants which belong to this same group. This group is the group of logical constants. The logical constants are those which constitute pure form; a formal proposition is a proposition which does not contain any other constants than logical constants. We have just reduced the deduction which proves that Socrates is mortal to the following form: "If *x* is a, then, if all the members of *a* are members of *b,* it follows that *x* is a *b*." The constants here are: is-*a, all, and if-then.* These are logical constants and evidently they are purely formal concepts.

Now, the validity of any valid deduction depends on its form, and its form is obtained by replacing the terms of the deduction by variables, until there do not remain any other constants than those of- logic. And conversely: every valid deduction can be obtained by starting from a deduction which operates on variables by means of logical constants, by attributing to variables definite values with which the hypothesis becomes true.

By means of this operation of generalization, we separate the strictly deductive element in an argument from the element which depends on the particularity of what is spoken about. Pure mathematics concerns itself exclusively with the deductive element. We obtain propositions of pure mathematics by a process of *purification*. If I say: "Here are two things, and here are two other things, therefore here are four things in all", I do not state a proposition of pure mathematics because here particular data come into question. The proposition that I have stated is an *application* of the general proposition: "Given any two things and also any two other things, there are four things in all." 'The latter proposition is a proposition of pure mathematics, while the former is a proposition of applied mathematics.

It is obvious that what depends on the particularity of the subject is the verification of the hypothesis, and this permits us to assert, not merely that the hypothesis implies the thesis, but that, since the hypothesis is true, the thesis is true also. This assertion is not made in pure mathematics. Here we content ourselves with the hypothetical form: It- any subject satisfies such and such a hypothesis, it will also satisfy such and such a thesis. It is thus that pure mathematics becomes entirely hypothetical, and concerns itself exclusively with any indeterminate subject, that is to say with a *variable.* Any valid deduction finds its form in a hypothetical proposition belonging to pure mathematics; but in pure mathematics itself we affirm neither the hypothesis nor the thesis, unless both can be expressed in terms of logical constants.

If it is asked why it is worthwhile to reduce deductions to such a form, I reply that there are two associated reasons for this. In the first place, it is a good thing to generalize any truth as much as possible; and, in the second place, an economy of work is brought about by making the deduction with an indeterminate x. When we reason al-out Socrates, we obtain results which apply only to Socrates, so that, if we wish to know something about Plato, we have to perform the reasoning all over again. But when we operate on x, we obtain results which we know to be valid for every x which satisfies the hypothesis. The usual scientific motives of economy and generalization lead us, then, to the theory of mathematical method which has just been sketched.

After what has just been said it is easy to see what must be thought about the intrinsic nature of propositions of pure mathematics. In pure mathematics we have never to discuss facts that are applicable to such and such an individual object; we need never know anything about the actual world. We are concerned exclusively with variables, that is to say, with any subject, about which hypotheses are made which may be fulfilled sometimes, but whose verification for such and such an object is only necessary for the *importance* of the deductions, and not for their truth. At first sight it might appear that everything would be arbitrary in such a science. But this is not so. It is necessary that the hypothesis truly implies the thesis. If we make the hypothesis that the hypothesis implies the thesis, we can only make deductions in the case when this new hypothesis truly implies the new thesis. Implication is a logical constant and cannot be dispensed with. Consequently we need true propositions about implication. If we took as premises propositions on implication which were not true, the consequences which would appear to flow from them would not be truly implied by the premises, so that we would not obtain even a hypothetical proof. This necessity for true premises emphasizes a distinction of the first importance, that is to say the distinction between a premise and a hypothesis. When we say "Socrates is a man, *therefore Soc*rates is mortal", the proposition "Socrates is a man" is a premise; but when we say: "*If* Socrates is a man, then Socrates is mortal", the proposition "Socrates is a man" is only a hypothesis. Similarly when I say: "If from p we deduce q and from q we deduce r, then from p we deduce r", the proposition "From p we deduce q and from q we deduce r" is a hypothesis, but the whole proposition is not a hypothesis, since I affirm it, and, in fact, it is true. This proposition is a rule of deduction, and the rules of deduction have a two-fold use in mathematics: both as premises and as a method of obtaining consequences of the premises. Now, if the rules of deduction were not true, the consequences that would be obtained by using them would not truly be consequences, so that we should not have even a correct deduction setting out from a false premise. It is this twofold use of the rules of deduction which differentiates the foundations of mathematics from the later parts. In the later parts, we use the same rules of deduction to deduce, but we no longer use them immediately as premises. Consequently, in the later parts, the immediate premises may be false without the deductions being logically incorrect, but, in the foundations, the deductions will be incorrect if the premises are not true. It is necessary to be clear about this point, for otherwise the part of arbitrariness and of hypothesis might appear greater than it is in reality.

Mathematics, therefore, is wholly composed of propositions which only contain variables and logical constants, that is to say, purely formal propositions-for the logical constants are those which constitute form. It is remarkable that we have the power of knowing such propositions. The consequences of the analysis of mathematical knowledge are not without interest for the theory of knowledge. In the first place it is to be remarked, in opposition to empirical theories, that mathematical knowledge needs premises which are not based on the data of sense. Every general proposition goes beyond the limits of knowledge obtained through the senses, which is wholly restricted to what is individual. If we say that the extension of the given case to the general is affected by means of induction, we are forced to admit that induction itself is not proved by means of experience. Whatever may be the exact formulation of the fundamental principle of induction, it is evident that in the first place this principle is general, and in the second place that it cannot, without a vicious circle, be itself demonstrated by induction.

It is to be supposed that the principle of induction can be formulated more or less in the following way. If we are given the fact that any two properties occur together in a certain number of cases, it is more probable that a new case which possesses one of these properties will possess the other than it would be if we had not such a datum. I do not say that this is a satisfactory formulation of the principle of induction; I only say that the principle of induction must be like this in so far as it must be an absolutely general principle which contains the notion of probability. Now it is evident that sense-experience cannot demonstrate such a principle, and cannot even make it probable; for it is only in virtue of the principle itself that the fact that it has often been successful gives grounds for the belief that it will probably be successful in the future. Hence inductive knowledge, like all knowledge which is obtained by reasoning, needs logical principles which are *a priori* and universal. By formulating the principle of induction, we transform every induction into a deduction; induction is nothing else than a deduction which uses a certain premise, namely the principle of induction.

In so far as it is primitive and undemonstrated, human knowledge is thus divided into two kinds: knowledge of particular facts, which alone allows us to affirm existence, and knowledge of logical truth, which alone allows us to reason about data. In science and in daily life the two kinds of knowledge are intermixed: the propositions which are affirmed are obtained from particular premises by means of logical principles. In pure perception we only find knowledge of particular facts: in pure mathematics, we only find knowledge of logical truths. In order that such knowledge be possible, it is necessary that there should be self-evident logical truths, that is to say, truths which are known without demonstration. These are the truths which are the premises of pure mathematics as well as of the deductive elements in every demonstration on any subject whatever.

It is, then, possible to make assertions, not only about cases which we have been able to observe, but about all actual or possible cases. The existence of assertions of this kind and their necessity for almost all pieces of knowledge which are said to be founded on experience shows that traditional empiricism is in error and that there is *a priori* and universal knowledge.

In spite of the fact that traditional empiricism is mistaken in its theory of knowledge, it must not be supposed that idealism is right. Idealism, at least every theory of knowledge which is derived from Kant, assumes that the universality of *a priori* truths comes from their property of expressing properties of the mind: I things appear to be thus because the nature of the appearance depends on the subject in the same way that, if we have blue spectacles, everything appears to be blue. The categories of Kant are the colored spectacles of the mind; truths *a priori* are the false appearances produced by those spectacles. Besides, we must know that everybody has spectacles of the same kind and that the color of the spectacles never changes. Kant did not deign to tell us how he knew this.

As soon as we take into account the consequences of Kant's hypothesis, it becomes evident that general and *a priori* truths must have the same objectivity, the same independence of the mind that the particular facts of the physical world possess. In fact, if general truths only express psychological facts, we could not know that they would be constant from moment to moment or from person to person, and we could never use them legitimately to deduce a fact from another fact, since they would not connect facts but our ideas about the facts. Logic and mathematics force us, then, to admit a kind of realism in the scholastic sense, that is to say, to admit that there is a world of universals and of truths which do not bear directly on such and such a particular existence. This world of universals must *subsist,* although it cannot *exist* in the same sense as that in which particular data exist. We have immediate knowledge of an indefinite number of propositions about universals: this is an ultimate fact, as ultimate as sensation is. Pure mathematics-which is usually called "logic" in its elementary parts-is the sum of everything that we can know, whether directly or by demonstration, about certain universals. On the subject of self-evident truths it is necessary to avoid a misunderstanding. Self-evidence is a psychological property and is therefore subjective and variable. It is essential to knowledge, since all knowledge must be either self-evident or deduced from self-evident knowledge. But the order of knowledge which is obtained by starting from what is self-evident is not the same thing as the order of logical deduction, and we must not suppose that when we give such and such premises for a deductive system, we are of opinion that these premises constitute what is self-evident in the system. In the first place self-evidence has degrees: It is quite possible that the consequences are more evident than the premises. In the second place it may happen that we are certain of the truth of many of the consequences, but that the premises only appear probable, and that their probability is due to the fact that true consequences flow from them. In such a case, what we can be certain of is that the premises imply all the true consequences that it was wished to place in the deductive system. This remark has an application to the foundations of mathematics, since many of the ultimate premises are intrinsically less evident than many of the consequences which are deduced from them. Besides, if we lay too much stress on the self-evidence of the premises of a deductive system, we may be led to mistake the part played by intuition (not spatial but logical) in mathematics. The question of the part of logical intuition is a psychological question and it is not necessary, when constructing a deductive system, to have an opinion on it.

To sum up, we have seen, in the first place, that mathematical logic has resolved the problems of infinity and continuity, and that it has made possible a solid philosophy of space, time, and motion. In the second place, we have seen that pure mathematics can be defined as the class of propositions which are expressed exclusively in terms of variables and logical constants, that is to say as the class of purely formal propositions. In the third place, we have seen that the possibility of mathematical knowledge refutes both empiricism and idealism, since it shows that human knowledge is not wholly deduced from facts of sense, but that *a priori* knowledge can by no means be explained in a subjective or psychological manner.

Chapter 6

OBJECTIVE KNOWLEDGE: A REALIST VIEW OF LOGIC, PHYSICS, AND HISTORY

Karl Popper

Man, some modern philosophers tell us, is alienated from his world: he is a stranger and afraid in a world he never made. Perhaps he is; yet so are animals, and even plants. They too were born, long ago, into a physico-chemical world, a world they never made. But although they did not make their world, these living things changed it beyond all recognition and, indeed, remade the small corner of the universe into which they were born. Perhaps the greatest of these changes was made by the plants. They radically transformed the chemical composition of the earth's whole atmosphere. Next in magnitude are perhaps the achievements of some marine animals which built coral reefs and islands and mountain ranges of limestone. Last come man, who for a long time did not change his environment in any remarkable way, apart from contributing, by deforestation, to the spread of the desert. Of course, he did build a few pyramids; but only during the last century or so, did he begin to compete with the reef-building corals. Still more recently he began to undo the work of the plants by slightly, though significantly, raising the carbon dioxide content of the atmosphere.

Thus, we have not made our world. So far we have not even changed it much, compared with the changes achieved by animals and plants. Yet we have created a new kind of product or artifact which promises in time to work changes in our corner of the world as great as those worked by our predecessors, the oxygen-producing plants, or the island-building corals. These new products, which are decidedly of our own making, are our myths, our ideas, and especially our scientific theories: theories about the world we live in.

I suggest that we may look upon these myths, these ideas and theories, as some of the most characteristic products of human activity. Like tools, they are organs evolving outside our skins. They are exosomatic artifacts. Thus we may count among these characteristic products especially what is called 'human knowledge'; where we take the word 'knowledge' in the objective or impersonal sense, in which it may be said to be contained in a book; or stored in a library; or taught in a university.

When referring to human knowledge, I shall usually have this objective sense of the word 'knowledge' in mind. This allows us to think of knowledge produced by men as analogous to the honey produced by bees: the honey is made by bees, stored by bees, and consumed by bees; and the individual bee which consumes honey will not, in general, consume only the bit it has produced itself: honey is also consumed by the drones which have not produced any at all (not to mention that stored treasure of honey which the bees may lose to bears or beekeepers). It is also interesting to note that, in order to keep up its powers to produce more honey, each working bee has to consume honey, some of it usually produced by other bees.

All this holds, by and large, with slight differences, for oxygen-producing plants and for theory-producing men: we, too, are not only producers but consumers of theories; and we have to consume other people's theories, and sometimes perhaps our own, if we are to go on producing.

'To consume' means here, first of all, 'to digest', as in the case of the bees. But it means more: our consumption of theories, whether those produced by other people or by us, also means criticizing them, changing them, and often even demolishing those, in order to replace them by better ones.

All these are operations which are necessary for the *growth of our knowledge;* and I again mean here, of course, knowledge in the objective sense.

I suggest that it looks at present as if it is this growth of human knowledge, the growth of our theories, which turns out, *human history* into a chapter so radically new in the history of the universe, and also in the history of life on earth.

All three of these histories-the history of the universe, the history of life on earth, and the history of man and of the growth of his knowledge-are, of course, themselves chapters of our knowledge. Consequently, the last of these chapters - that is, the history of knowledge-will consist of knowledge about knowledge. It will have to contain, at least implicitly, theories about theories, and especially theories about the way in which theories grow.

I shall, therefore, before going any further into my topic, present a general tetradic schema which I have found more and more useful as a description of the growth of theories. It is as follows:

$$P_1 \gg TT \gg EE \gg P_2.$$

Here '*P*' stands for 'problem'; '*TT*' stands for 'tentative theory'; and '*EE*' stands for '(attempted) error-elimination', especially by way of critical discussion. My tetradic schema is an attempt to show that the result of criticism, or of error-elimination, applied to a tentative theory, is as a rule the emergence of a new problem; or, indeed, of several new problems. Problems, after they have been solved and their solutions properly examined, tend to beget problem-children: new problems, often of greater depth and ever greater fertility than the old ones. 'This can be seen especially in the physical sciences; and I suggest that we can best gauge the progress made in any science by the distance in depth and expectedness between P_1 and P_2: the best tentative theories (and all theories are tentative are those which give rise to the deepest and most unexpected problems.

My tetradic schema can be elaborated in various ways; for example, by writing it as follows:

$$\gg TT_a \gg EE_a \gg P_{2a}$$
$$P_1 \gg TT_b \gg EE_b \gg P_{2b}$$
$$\gg TT_n \gg EE_n \gg P_{2n}$$

In this form the schema would indicate that, if we can, we should propose many theories as attempts to solve some given problem, and that we should critically examine each of our tentative solutions. We then find that each gives rise to new problems; and we may follow up those which promise the most novel and most interesting new problem: if the new problem, P_{2b}, say, turns out to be merely the old *P*, in disguise, then we say that our theory only manages to *shift the problem* a little; and in some cases we may take this as a decisive objection to the tentative theory, TT_b.

This shows that error-elimination is only *part* of our critical discussion: our critical discussion of the competing tentative theories may compare them, and assess them, from many different points of view. The decisive point is, *of course,* always: how well does our theory solve its problems; that is, P_1.

At any rate, one of the things we wish to achieve is to learn something new. According to our schema, progressiveness is one of the things we demand of a good tentative theory: and it is brought out by the critical discussion of it: the theory is progressive if our discussion shows that *it has really made a deference to the problem we wanted to solve;* that is, if the newly emerging problems are different from the old ones.

If the newly emerging problems are different then we can hope to learn a great many new things when we proceed to solve them in turn.

Thus my tetradic schema can be used to describe the emergence of new problems and, consequently, the emergence of new solutions-that is, new theories; and I even want to present it as an attempt to make sense of the admittedly vague idea of emergence-as an attempt to speak of emergence in a rational manner. I should like to mention that it can be applied not only to the emergence of new scientific problems and, consequently, new scientific theories, but to the emergence of new forms of behavior, and even new forms of living organisms.

Let me give you an example. P_1 may be, say, a certain problem concerning the survival of a species, such as the problem of reproduction, of producing offspring. According to Darwin, this survival problem has found a good solution if the species survives; any other tentative solution will be eliminated by the disappearance of both the solution and the species.

According to my schema, the attempted error-elimination, that is, the struggle for survival will bring out the inherent weakness of each of the proposed solutions in the form of a *new problem.* For example, the new problem may be that the parent organisms and their offspring are threatening to suffocate one another. This new problem may in turn be solved; for example, the organisms may develop a method of scattering or disseminating their offspring; or else the new problem may be solved by the establishment of a common economy, comprising several organisms. Perhaps the transition from unicellular to multicellular organisms proceeded in this way.

However this may be, my schema shows that there may be more than Darwin's alternative, *'survive or perish',* inherent in the process of error-elimination: error-elimination may bring out new emerging problems, specifically related to the old problem and to the tentative solution.

In what follows I shall use my schema, sometimes only implicitly; and I shall refer to emergence, assuming that my schema makes this idea sufficiently respectable within what I hope will be a rational discussion. I propose to deal with some aspects of the growth of knowledge under four headings:

1. Realism and Pluralism: Reduction versus Emergence.
2. Pluralism and Emergence in History.
3. Realism and Subjectivism in Physics.
4. Realism in Logic.

1. Realism and Pluralism: Reduction Versus Emergence

Man produces not only scientific theories but many other ideas - for example, religious or poetical myths or, say, plots for stories.

What is the characteristic difference between a scientific theory and a work of fiction? It is not, I hold, that the theory is possibly true while the descriptions in the story are not true, although truth and falsity have something to do with it. The difference is, I suggest, that the theory and the story are embedded in different critical traditions. They are meant to be judged by quite different traditional standards (even though these standards may have something in common).

What characterizes the theory is that it is offered as a solution to a scientific problem; that is, either a problem that has arisen before, in the critical discussion of earlier tentative theories, or (perhaps) a problem discovered by the author of the theory now offered, but discovered within the realm of the problems I and solutions belonging to the scientific tradition.

However, I am not leaving it at that. For the scientific tradition in its turn is, or was until recently, characterized by what may be called *scientific realism*. That is to say, it was inspired by the ideal of finding *true solutions* to its problems: solutions which corresponded to the facts.

This regulative ideal of finding theories which correspond to the facts is what makes the scientific tradition a realist tradition: it distinguishes between the world of our theories and the world of facts to which these theories belong.

Moreover, the natural sciences with their critical methods of problem solving, and some of the social sciences too, especially history and economics, have represented for quite a long time our best efforts in problem solving and fact finding (by fact finding I mean, of course, the discovery of statements or theories which correspond to facts). Thus these sciences contain, by and large, the best statements and theories from the point of view of truth; that is, those giving the best description of the world of facts, or of what one calls 'reality'.

Now let us look at certain relations that hold between some of these sciences.

Take physics and chemistry for example; sciences which make assertion about all physical things and physical states, including living organisms.

Physics and chemistry are not very different, and there seems to be no great difference in the kind of things to which they apply, except that chemistry, as it is usually understood, becomes inapplicable at very high temperatures and also, perhaps, at very low ones. It therefore would not be very surprising if the hopes, held for a long time, that chemistry can be reduced to physics, were to come true, as indeed they seem to be doing.

Here we have a real paradigm case of a *'reduction';* by a *reduction* I mean, of course, that all the findings of chemistry can be fully explained by (that is to say, deduced from) the principles of physics.

Although such a reduction would not be very surprising, it would be a very great scientific success. It would not only be an exercise in unification, but a real advance in understanding the world.

Let us assume that this reduction has been carried out completely. This might give us some hope that we may also reduce one day all the biological sciences to physics.

Now this would be a spectacular success, far greater than the reduction of chemistry to physics. Why? Because the kind of things to which physics and chemistry apply are really very similar from the start. Only think how difficult it would be to say whether the atomic theory is a physical or a chemical theory. In fact, for a long time it was both; and it is this common bond which provides the link which may lead, or perhaps has led, to their unification.

With living organisms the situation is different. They are, no doubt, subject to all kinds of physical and biological laws. Yet there appears to be some prima facie difference between living organisms and non-living things. Admittedly, we learn from science that there are transitory or intermediate stages, and also intermediate systems; and this gives us hope that a reduction might be achieved one day. Moreover, it seems not at all improbable that recent tentative theories about the origin of life on earth might be successfully put to the test, and that we might be able to create primitive living organisms artificially.

But even this would not necessarily mean a complete reduction. This is shown by the fact that chemists were able to create all sorts of chemicals, inorganic and organic, before understanding even their chemical composition, to say nothing about their physical structure. Thus even the control of chemical processes by purely physical means is not as such equivalent to a reduction of chemistry to physics. Reduction means much more. It means *theoretical* understanding: the *theoretical* penetration of the new field by the old field.

Thus we might find a recipe for creating some primitive forms of life from non-living matter without understanding, theoretically, what we were doing. Admittedly, this would be a tremendous encouragement to all those who seek for a reduction, and rightly so. But the way to a reduction might still be long; and we could not know whether it was not even impassable: there may be no theoretical reduction of biology to physics, just as there seems to be neither a theoretical reduction of mechanics to electrodynamics, nor a theoretical reduction the other way round.

If the situation is such that, on the one hand, living organisms may originate by a natural process from non-living systems, and that, on the other hand, there is no complete theoretical understanding of life possible in physical terms, and then we might speak of life as an *emergent* property of physical bodies, or of matter.

Now I want to make it quite clear that as a rationalist I wish and hope to understand the world and that I wish and hope for a reduction. At the same time, I think it quite likely that there may be no reduction possible; it is conceivable that life is an *emergent* property of physical bodies.

My point here is that those believers in reduction who, for some philosophical or other reason, adopt *a priori* the dogmatic position that reduction must be possible, in a way destroy their triumph should reduction ever be achieved. For what will then be achieved ought to have been achieved all the time; so their triumph will be only the uninteresting one of having been proved right by events.

Only those who assert that the question cannot be settled *a priori* can claim that any successful reduction would be a tremendous discovery.

I have dwelt on this point so long because it has some bearing on the position of the next rung of the ladder-the emergence of consciousness.

There are philosophers, called 'radical behaviorists' or 'physicalists', who think that they have *a priori* reasons, such as *Ockham's razor,* for asserting that our introspection of mental states or events, and our reports about mental states or events, are simply introspections and reports about ourselves qua physical systems: they are reports about physical states of these systems.

Two philosophers expected here this morning have defended such a view with brilliant arguments. They are Herbert Feigl and Willard Van Orman Quine. I should like to make a few critical remarks about their views.

Quine says, with a reference to Carnap and Feigl, that if theoretical progress can be 'achieved by ... positing distinctive mental states ... behind physical behavior, surely as much ... could be achieved by positing..... certain correlative physiological states and events instead.... Lack of a detailed physiological explanation of the states is scarcely an objection to acknowledging them as states of human bodies.... The bodily states exist anyway; why add the others?'

Let me point out that Quine speaks here as a realist: 'The bodily states exist anyway', he says. Nevertheless, from the point of view I am adopting here, he is not what I should call a 'scientific realist': he does not wait to see whether science will achieve a reduction here, as perhaps it may one day; instead he applies Ockham's razor, pointing out that mental *entities* are not necessary for the theory.

But who knows what Ockham or anybody else might mean here by necessity? If mental entities or, better, mental states should exist - and I myself do not doubt that they do exist, then positing mental states is necessary for any true explanation of them; and should they one day be reduced to physical states, hen this will be a tremendous success. But there will be no success at all if we reject their existence by merely noting that we can explain things without them, by the simple method of confining ourselves to physical things and their behavior.

To sum up my argument in brief: philosophical speculations of a materialistic or physicality character are very interesting, and may even be able to point the way to a successful scientific education. But they should be frankly tentative theories (as think Feigl's theories are). Some physicalists do not, however, consider their theories as tentative, but as proposals to express everything in a physicalistic language; and they think these proposals have much in their favor because they are undoubtedly convenient: inconvenient problems such as the body-mind problem do indeed, most conveniently, disappear. So these physicalists think that there can be no doubt that these problems should be eliminated as pseudo-problems.

To this I would reply that by the same method we could have eliminated *a priori* all chemical states and problems connected with them: we could have said that they were obviously physical, and that there was no need to specify them in detail: that all we needed to do was to postulate the existence of some physical state correlative to each chemical state. : I think it is clear that the general adoption of such a proposal would have led to the attitude of not looking for the detailed reduction of chemistry to physics. No doubt, it would have dissolved the analogue of the body-mind problem-the problem of the relation of physics to chemistry; but the solution would have been linguistic; and as a consequence we should not have learned anything about the real world.

All this leads me to assert that realism should be at least tentatively pluralistic, and that realists should subscribe to the following pluralistic postulate:

> We must beware of solving, or dissolving, factual problems linguistically; that is, by the all too simple method of refusing to talk about them. On the contrary, we must be pluralists, at least to start with: we should first emphasize the difficulties, even if they look insoluble, as the body-mind problem may look to some.
>
> If we can then reduce or eliminate some entities by way of scientific reduction, let us do so by all means, and be proud of the gain in understanding.
>
> So I would say: let us work out in every case the arguments for emergence in detail, *at any rate before attempting reduction.*

To sum up and sharpen the considerations advanced in this section:

The reduction of chemistry to physics, apparently now well on the way, may be described as a paradigm case of a genuine scientific reduction which satisfies all the requirements of a good scientific explanation.

'Good' or 'scientific' reduction is a process in which we learn much that is of great importance: we learn to understand and to explain the theories about the field to be reduced (in this case chemistry) and we learn a great deal about the power of the reducing theories (in this case physics).

It is conceivable, although not yet certain that the reduction of chemistry to physics will be completely successful. It is also conceivable, though less likely, that we may one day have *good reductions* of biology, including physiology, to physics, and of psychology to physiology, and thus to physics.

I call *bad reduction or ad hoc* reduction the method of reduction by merely linguistic devices; for example, the method of physicalism which suggests that we postulate *ad hoc* the existence of physiological states to explain behavior which we previously explained by postulating (though not by postulating *ad hoc)* mental states. Or in other words, by the linguistic device of saying that I report on a *physiological* state of mine when I report that I now feel that I understand the Schrödinger equation.

This second kind of reduction or the use of Ockham's razor is bad, because it prevents us from seeing the problem. In, the picturesque as well as hard-hitting terminology of Imre Lakatos, it is a disastrous case of a *'degenerating problem shift'*; and it may prevent either a good reduction, or the study of emergence, or both.

In order to avoid this disastrous method we must in each case try to learn as much as possible about the field which we hope to reduce. It may be that the field resists reduction; and we may even possess arguments to show why the field cannot be reduced. In this case we may have an example of genuine emergence.

If I may perhaps end my comments on the degenerating of behaviorism (especially linguistic behaviorism) with the following remark.

Behaviorists and materialists are anti-idealists: and they are, rightly, opponents of Berkeley's *'esse = percipi'* or

to be = to be observable.

According to them, 'to be' is 'to be material', 'to behave as a body in space and time'. Nevertheless, it may be said that they do adhere, unconsciously, to Berkeley's equation, although they put it in a slightly different verbal form:

to be = to be observed

or perhaps

to be = to be perceived.

For they say that only those things exist which can be observed. They do not realize that *all observation involves interpretation in the light of theories,* and that what they call 'observable' is what is observable in the light of pretty old-fashioned and primitive theories. Though I am all for common sense, I am also for enlarging the realm of common sense by learning from science. At any rate, *it is not science but dubious philosophy (or outdated science) which leads to idealism, phenomenalism, and positivism; or to materialism and behaviorism,* or to any other of anti-pluralism.

2. Pluralism and Emergence in History

I shall not speak about the history of the universe, but only say a few words about the history of life on earth.

It seems that a very promising start has recently been made towards reconstructing the conditions under which life *emerged* on earth; and I think we may, perhaps, expect some major success soon. But while sanguine about emergence, even experimental emergence, I feel very skeptically inclined about reduction. This is due to certain thoughts of mine about the evolution of life.

It seems to me that evolutionary processes or major evolutionary changes are as unpredictable as historical processes or major historical changes. I hold this view because I am strongly inclined towards an indeterminist view of the world, somewhat more radical than Heisenberg's: my indeterminism includes the thesis that even classical physics is indeterministic, and is thus more like that of Charles Sanders Peirce, or that of Alfred Landé. And I think that evolution proceeds largely probabilistically, under constantly changing conditions or problem situations, and that every tentative solution, whether more successful or less successful or even completely unsuccessful, creates a new problem situation. This seems to me to prevent a complete reduction as well as a complete understanding of the processes of life, although it does not prevent constant and far-reaching progress towards such understanding. (This argument should not be taken to be like Bohr's application of his idea of complementarity to living organisms - an argument which seems to me very weak indeed.)

But I want to speak in this section mainly about human history, about the story of mankind. This, as I have indicated, is very largely the history of our knowledge - of our theories about the world - and, of course, of the repercussions of these products, which are of our own making, upon ourselves and our further productions.

It is obvious that one can adopt a physicalist or materialist attitude towards these theoretical products of ours; and it might be suspected that my emphasis upon the objective sense of knowledge - my emphasis upon theories as contained in books collected in libraries and as taught in universities-indicates that I sympathize with the physicalist or materialist interpretation of theories; I mean an interpretation which sees language as consisting of physical objects - noises, or printed letters-and which sees ourselves as conditioned, or dispositional, to react to these noises or letters with certain characteristic kinds of physical behavior.

But nothing is further from my intention than to encourage *ad hoc* reductions of this kind. Admittedly, if forced to choose between any subjectivist or personalist view of human knowledge and the materialist or physicalist view I have just tried to sketch, I should choose the latter; but this is emphatically *not* the alternative.

The history of ideas teaches us very clearly that ideas emerge in logical or, if the term is preferred, in dialectical contexts. My various schemata such as

$$P_1 \gg TT \gg EE \gg P_2$$

may indeed be looked upon as improvements and rationalizations of the Hegelian dialectical schema: they are rationalizations because they operate entirely within the classical logical organon of rational criticism, which is based upon the so-called law of contradiction; that is to say, upon the demand that contradictions, whenever we discover them, must be eliminated. Critical error-elimination on the scientific level proceeds by way of a conscious search for contradictions.

Thus history, and especially the history of ideas, teaches us that if we want to understand history, we must understand ideas and their objective logical (or dialectical) relationships.

I do not believe that anybody who has ever seriously gone into any chapter of the history of ideas will think that a reduction of these ideas could ever be successful. But I take it as my task here not so much to argue against the possibility of any reduction, as to argue for the recognition of emergent entities, and for the need to recognize and describe these emergent entities before one can seriously think about their possible elimination by way of reduction.

One of my main arguments for the emergent character of theories I have given elsewhere. My argument depends upon the conjecture that there is such a thing as a genuine growth of scientific knowledge; or in practical terms, that tomorrow, or a year hence, we may propose and test important theories of which nobody has seriously thought so far. If there is growth of knowledge in this sense, then it cannot be predictable by scientific means. For he who could so predict today by scientific means our discoveries of tomorrow could make them today; which would mean that there would be an end to the growth of knowledge.

On the other hand, unpredictability in principle has always been considered as the salient point of emergence; and it seems to me that my argument shows at any rate that the growth of knowledge must be unpredictable in principle.

But there are other arguments for the emergent character of theories, or of knowledge in the objective sense. I shall only mention an argument or two against the very popular and very naive view that theories can be reduced to the mental states of those who produce them, or of those who understand them. (Whether or not these mental states themselves may then perhaps be reduced to physical states in turn will not be further discussed.)

The idea that a theory in its objective or logical sense may be reduced to the mental states of those who hold the theory takes, as a rule, the form that the theory just *is* a thought. But this is a trivial mistake: it is the failure to distinguish between two senses of the word 'thought'. In its subjective sense, the word 'thought' describes a mental experience or a mental process. But two mental experiences or processes, though they may stand in causal relations to each other, cannot stand in logical relations to each other.

Thus, if I say that certain ideas of the Buddha agree with certain ideas of Schopenhauer, or that they contradict certain ideas of Nietzsche, then I am not speaking about the mental thought-processes of these people or about their interrelations. If I say, however, that Nietzsche was influenced by certain ideas of Schopenhauer, then I do mean that certain thought processes of Nietzsche's were causally influenced by his reading of Schopenhauer. So we have actually these two different worlds, the world of *thought-processes,* and the world of the *products* of thought-processes. While the former may stand in *causal* relationships, the latter stand in *logical* relationships.

The fact that certain theories are incompatible is a logical fact, and holds quite independently of whether or not anybody has noticed or understood this incompatibility. These purely objective logical relationships are characteristic of the entities which I have called theories, or knowledge, in the objective sense.

This may also be seen from the fact that the person who produces a theory may very often not understand it. Thus it might be argued without paradox that Erwin Schrödinger did not fully understand the Schrödinger equation, at any rate not until Max Born gave his statistical interpretation of it; or that Kepler's area law was not properly understood by Kepler, who seems to have disliked it.

In fact, understanding a theory is something like an infinite task, so that we may well say that a theory is never fully understood, even though some people may understand some theories extremely well. Understanding a theory has, indeed, much in common with understanding a human personality. We may know or understand a man's system of dispositions pretty well; that is to say, we may be able to predict how he would act in a number of different situations. But since there are infinitely many possible situations, of infinite variety, a full understanding of a man's dispositions does not seem to be possible. Theories are similar: a full understanding of a theory would mean understanding all its logical consequences. But these are infinite in a non-trivial sense: there are infinitely many situations of infinite variety to which the theory might be applicable; that is to say, upon which some of its logical consequences may bear; and many of these situations have never been thought of; their possibility may not yet have been discovered. But this means that nobody, neither its creator nor anybody who has tried to grasp it, can have a full understanding of all the possibilities inherent in a theory; which shows again that the theory, in its logical sense, is something objective and something objectively existing - an object that we can study, something that we try to grasp. It is no more paradoxical to say that theories or ideas are our products and yet not fully understood by us than to say that our children are our products and yet not fully understood by us, or that honey is a product of the bee, yet not fully understood by any bee.

Thus, the study of the history of our theories or ideas-and a good case could be made for the view that all human history is largely a history of our theories or ideas-should make us all pluralists. For what exist, for the historian, are people in physical, social, mental, and ideological problem situations; people producing ideas by which they try to solve these problems, ideas which they try to grasp, to criticize, to develop.

The student of the history of ideas will find that ideas have a kind of life (this is a metaphor, of course); that they can be misunderstood, rejected, and forgotten; that they can reassert themselves, and come to life again. Without metaphor, however, we can say that they are not identical with any man's thought, or belief; that they can exist even if universally misunderstood, and rejected.

All this may be reminiscent of Plato and Hegel. But there are great differences here. Plato's 'ideas' were eternal, unchanging conceptions or notions; Hegel's were dialectically self-changing conceptions or notions. The ideas which I find most important are not conceptions or notions at all. They correspond not to words but to statements or propositions.

In opposition to Plato and Hegel I consider *tentative theories* about the world-that is, hypotheses together with their logical consequences-as the most important citizens of the world of ideas; and I do not think (as Plato did) that their strangely non-temporal character makes them eternal and thereby *more real* than things that are generated and are subject to change, and to decay. On the contrary, a thing that can change and perish should for this very reason be accepted as prima facie real; and even an illusion is, *qua* illusion, a real illusion.

This is important in connection with the problem of time, and of change.

A historian cannot, I think, accept the doctrine that time and change are illusions; a doctrine upheld by some great physicists and philosophers such as Parmenides, Weyl, and Schrödinger. Nothing is more real than an event, an occurrence; and every event involves some change.

That the pluralistic universe in which the historian lives, with its individual men living individual lives, trying to solve their problems, producing children, and ideas about them, hoping and fearing and deceiving themselves and others, but always theorizing, and often seeking not only happiness but also truth-that this pluralistic universe should be successfully 'reduced' to one or another kind of monism-this seems to me not only unlikely, but impossible. But this is not my point here. My point is that only after recognizing the plurality of what there is in this world can we seriously begin to apply Ockham's razor. To invert a beautiful formulation of Quine's, only if Plato's beard is sufficiently tough, and tangled by many entities, can it be worth our while to use Ockham's razor. That the razor's edge will be dulled in being used for this tough job is only to be expected. The job will no doubt be painful. But it is all in a day's work.

3. Realism and Subjectivism in Physics

There are two important fields in modern physics in which physicists have allowed subjectivism not only to enter, but to play an essential role: Boltzmann's theory of the subjectivity of the direction of time, and Heisenberg's interpretation of the indeterminacy formulae as determining a lower limit to the effect of the observer's interference with the observed object.

There was also another intrusion of the subject, or of the observer, w hen Einstein brought in the observer in a number of imaginary thought experiments intended to elucidate relativity; but this is a field from which the observer was exorcised, slowly but steadily, by Einstein himself.

I shall not discuss this point further, nor shall I discuss the subjective theory of time which, in trying to tell us that time and change are human illusions, forgets that they are very real illusions which have in no way been reduced to anything else (and which, I conjecture, are not amenable to reduction). I shall not discuss all this because I have done so only recently. I merely want to say a few words about the Heisenberg formulae and their interpretation.

These formulae are usually derived in a fairly complicated manner; there is, for example, an interesting derivation by Weyl and another rather complicated by Born.

Yet in fact the Heisenberg formula for energy depends neither on wave mechanics nor on Heisenberg's matrix mechanics; nor do we need the commutation relations (which according to Hills are insufficient for the derivation of the formulae). It simply does not depend on the revolutionary new quantum mechanics of 1925-6, but follows directly from Planck's old quantum postulate of 1900:

$E = hf.$

From this we get immediately

(2) $\Delta E = \mathrm{h}\, \Delta f.$

By using the principle of harmonic resolving power,

(3) Δf approx $= 1/\Delta t$,

we obtain from (2) and (3)

(4) ΔE approx $= \mathrm{h} / \Delta t$,

this leads at once to

(5) $\Delta E\ \Delta t$ approx $= h$;

that is to say, a form of Heisenberg's so-called *indeterminacy formulae.*

In precisely the same way we obtain the Heisenberg formula for position and momentum from Duane's principle (whose analogy to Planck's principle has recently been stressed by Alfred Landé). It may be written

(6) Δp_i approx $= h / \Delta q_i$

According to Landé this may be interpreted as follows: a body (such as a grid or a crystal) endowed with the space-periodicity Δqi is entitled to change its momentum pi in multiples of Δp_i approx $= h / \Delta q_i$.

From (6) we obtain at once

(7) $\Delta p_i \, . \, \Delta q_i$ approx $= h$,

this is another form of Heisenberg's indeterminacy formulae.

Considering that Planck's theory is a statistical theory, the Heisenberg formulae can be most naturally interpreted as statistical *scatter relations,* as I proposed more than thirty years ago. That is, they say nothing about the possible precision of measurements, nor anything about limits to our knowledge. But if they are scatter relations, they tell us something about the limits to the homogeneity of quantum-physical states, and therefore, though indirectly, about predictability.

For example, the formula $\Delta p_i \, . \, \Delta q_i$ approx $= h$ (which can be obtained from Duane's principle just as $\Delta E \, . \, \Delta T$ approx $= h$ can be obtained from Planck's principle) tells us, simply, that if we determine the coordinate x of a system (say, an electron) then, upon repetition of the experiment, the momentum will scatter.

Now how can such an assertion are tested? By making a long series of experiments with a fixed shutter opening Δx and by measuring, in every single case, the momentum P_x. If these moments scatter as predicted, then the formula has survived the test. But this shows that in order to test the scatter relations, we have actually measured, in every case, p_x with a precision far greater than Δp_x; for otherwise we could not speak of Δp_x, as the scatter of p_x.

Experiments of the kind described are carried out every day in all physical laboratories. But they refute Heisenberg's indeterminacy interpretation, since measurements (though not the predictions based upon them) are more precise than this interpretation permits.

Heisenberg himself noted that such measurements are possible, but he said that it was 'a matter of personal belief' or personal taste' whether or not we attach any meaning to them; and ever since this remark they have been universally disregarded as meaningless. But they are not meaningless, for they have a definite function: they are tests of the very formulae in question; that is, of the indeterminacy formulae qua scatter relations.

There is, therefore, no reason whatever to accept either Heisenberg's or Bohr's subjectivist interpretation of quantum mechanics. Quantum mechanics is a statistical theory because the problems it tries to solve-spectral intensities, for example -are statistical problems. There is, therefore, no need here for any philosophical defense of its non-causal character.

The irreducibility of statistical theories to deterministic theories (rather than the incompatibility of these two kinds of theories) should, however, be established. Arguments to this effect have been offered by Landé, and very different ones by myself.

To sum up, there is no reason whatsoever to doubt the realistic and objectivistic character of all physics. The role played by the observing subject in modern physics is in no way different from the role he played in Newton's dynamics or in Maxwell's theory of the electric field: the observer is, essentially, the man who tests the theory. For this, he needs a lot of other theories, competing theories and auxiliary theories. All this shows that we are not so much observers as thinkers.

4. Realism in Logic

I am opposed to looking upon logic as a kind of game. I know about so-called alternative systems of logic and I have actually invented one myself, but alternative systems of logic can be discussed from very different points of view. One might think that it is a matter of choice or convention which logic one adopts. I disagree with this view.

My theory is briefly this. I look upon logic as the theory of deduction or of derivability, or whatever one chooses to call it. Derivability or deduction involves, essentially, *the transmission of truth and the retransmission of falsity:* in a valid inference truth is transmitted from the premises to the conclusion. This can be used especially in so-called 'proofs'. But falsity is also retransmitted from the conclusion to (at least) one of the premises, and this is used in disproof or reputations, and especially in *critical discussions.*

We have premises and a conclusion; and if we show that the conclusion is false, and assume that the inference is valid, we know that at least one of our premises must be false. This is how logic is constantly used in critical discussion, for in a critical discussion we attempt to show that something is not in order with some assertion. We attempt to show it; and we may not succeed: criticism may be validly answered by counter-criticism.

What I should wish to assert is (1) that criticism is a most important methodological device; and (2) that if you answer criticism by saying, 'I do not like your logic: your logic may be all right for you, but I prefer a different logic, and according to my logic this criticism is not valid', then you may undermine the method of critical discussion.

Now I should distinguish between two main uses of logic, namely (1) its use in the demonstrative sciences-that is to say, the mathematical sciences-and (2) its use in the empirical sciences.

In the demonstrative sciences logic is used in the main for proofs-for the transmission of truth-while in the empirical sciences it is almost exclusively used critically-for the retransmission of falsity. Of course, applied mathematics comes in too, in which we implicitly make use of the proofs of pure mathematics, but the role of mathematics in the empirical sciences is somewhat dubious in several respects. (There exists a wonderful article by Schwartz to this effect.)

Thus in the empirical sciences logic is mainly used for criticism; that is, for refutation (Remember my schema P_1 » *TT* » *EE*» P_2.)

Now, what I wish to assert is this. If we want to use logic in a critical context, then we should use a very strong logic, the strongest logic, so to speak, which is at our disposal; for we want our criticism to be *severe.* In order that the criticism should be severe we must use the full apparatus; we must use all the guns we have. Every shot is important. It doesn't matter if we are over-critical: if we are, we shall be answered by counter-criticism.

Thus we should (in the empirical sciences) use the full or classical or two-valued logic. If we do not use it but retreat into the use of some weaker logic - say, the intuitionist logic, or some three-valued logic (as Reichenbach suggested in connection with quantum theory)-then, I assert, we are not critical enough; it is a sign that something is rotten in the state of Denmark (which in this case is the quantum theory in its Copenhagen interpretation, as I indicated earlier).

Now let us look, by contrast, at proofs. Every mathematician knows that considerable interest lies in proving a theorem with the help of a *minimum apparatus.* A proof which uses stronger means than necessary is mathematically unsatisfactory, and it is always interesting to find the weakest assumptions or minimum means which have to be used in a proof. In other words, we want the proof not only to be sufficient - that is to say valid-but we want it if possible to be necessary, in the sense that a minimum of assumptions have been used in the proof. This, I admit, is a somewhat sophisticated view. In unsophisticated mathematics we are happy and grateful if we can prove anything, but in more sophisticated mathematics we really want to know what is *necessary* for proving a theorem.

So if one can prove mathematical theorems with methods weaker than the full battery of classical logic, then this is extremely interesting from a mathematical point of view. Thus in proof theory we are interested in weakening if possible our classical logic, and we can, for example, introduce intuitionist logic or some other weaker logic such as positive logic, and investigate how far we can get without using the whole battery.

I think, incidentally, that the term 'intuitionist logic' is a misnomer. It is just a name for a very interesting and somewhat weakened form of classical logic invented by Brouwer and formalized by Heyting. I certainly do not want to say anything in favor of the philosophical theory called intuitionism though I should like to say something in favor of the Brouwer-Heyting logic. But I trust it will not be supposed that I am in any sense defending the authority of intuition in philosophy or logic or anywhere else. Leaving aside for the moment Brouwerian logic, one might say that intuitionism is the doctrine that intuitions are not only important but generally *reliable.* As against this I think that intuitions are very important but that as a rule they do not stand up to criticism. So I am not an intuitionist. However, Brouwerian or so-called 'intuitionist logic' is, from the standpoint of the present discussion, important because it is just a part, a genuine part, and thus a weakened form, of classical logic; that is to say, every inference which is valid from the point of view of intuitionist logic is also valid from the point of view of classical logic, while the opposite s not the case: we have inferences which may be validly drawn in classical logic but which are not valid in intuitionist logic. Thus if I can prove a theorem (so far proved only by classical means) with intuitionist logic, I have made a real mathematical discovery; for mathematical discoveries do not consist only in finding new proofs of new theorems, but they consist also in finding new proofs of old theorems; and a new proof of a theorem will be especially interesting if it uses weaker means than the old proof. A proof using stronger means one can always have for the asking, *a fortiori;* yet finding a weaker proof is a real mathematical achievement.

So intuitionistic logic is a very interesting approach to mathematics because it tries to prove as many mathematical theorems as possible with reduced logical means.

Intuitionistic logic has a further advantage: one can show that in it the so-called 'law of excluded middle' is not demonstrable (although it is a well-formed formula of the system) One can also show that if in any system whatsoever some well-formed formula is not demonstrable, then the system must be consistent. Generally speaking, the weaker the logical means we use, the less is the danger of inconsistency the danger that a contradiction is derivable. So intuitionist logic can also be looked upon as an attempt to make more certain that our arguments are consistent and that we do not get into hidden inconsistencies or paradoxes or antinomies. How safe such a weakened logic is, as such, is a question into which I do not want to enter now; but obviously it is at least a little safer than the full classical logic. I do not suppose it is always safe, but that is not my point. My point is this. If you wish to prove, or to establish something, you should use weak means. But for disestablishing it-that is to say, for criticizing it-we may use strong means. Of course someone might say, 'Look here, I can refute you even with weak means; I do not even need to use the whole of intuitionist logic.' Still, that is not very important. The main thing is that for the rationalist any criticism is welcome-though he may reply to it by criticizing the criticism.

Now this rationalist view is a realist view of logic. First, because it looks upon logic partly in connection with the methodology of the natural sciences which, I have tried to argue, is a realistic affair. Secondly, and this is a very special point, because it looks upon logical inference as truth transmitting or falsity re-transmitting; that is to say, it is concerned with the idea of truth.

I would assert that not the least important of the achievements of Alfred Tarski is that by introducing two ideas into logic, he has actually made logic very much a realistic affair. The first is Tarski's idea (partly anticipated by Bolzano) that logical consequence is truth transmission. The second, I would say, is the rehabilitation of the correspondence theory of truth, the rehabilitation of the idea that truth is simply correspondence with the facts.

I think I may differ here a little from Quine, because I think that this idea of Tarski's ought to be interpreted as destructive of relativism, and because I think that Tarski's claim that his theory of truth is an 'absolutistic' theory of truth is correct. In order to explain this point, I will recount a very old story with a slightly new point to it. The old story is the story of the three main theories of truth. The new point is the elimination of the word 'truth' from the story, and with it, of the appearance that we are dealing here with words, or verbal definitions. However, for this elimination some preparatory discussion is needed.

Of the three main theories of truth, the oldest was the correspondence theory, the theory that truth is correspondence with the facts, or to put it more precisely, that a statement is true if (and only if) it corresponds to the facts, or if it adequately describes the facts. This is the theory which I think Tarski has rehabilitated. The second theory is the so-called coherence theory: a statement is regarded as true if (and only if) it coheres with the rest of our knowledge. The third theory is that truth is pragmatic utility or pragmatic usefulness.

Now, the coherence theory has all sorts of versions of which I shall mention just two. According to the first, truth is coherence with our beliefs, or more precisely, a given statement is true if it coheres with the rest of our beliefs. This I find a bit disconcerting because I do not want to put beliefs into logic, for well-known reasons. (If Peter believes *p,* and if *p* and *q* are interdeducible, we might say that Peter is logically bound to believe *q*. Yet he may not know that *p* and *q* are interdeducible, and he may in fact disbelieve *q*.)

According to the second version of the coherence theory a certain given statement, of which we do not know whether it is true or not, is to be accepted as true if (and only if) it coheres with the statements we have previously accepted.

This version has the effect of making our knowledge utterly conservative: 'entrenched' knowledge can hardly be overthrown.

The theory of pragmatic utility is especially concerned with the problem of theories in the natural sciences such as physics. It says that we should accept a physical theory as true if it turns out in tests, and other applications, to be pragmatically useful, or successful.

I propose now to use something like a trick. My trick consists in this. I shall very soon, until very near the end of this paper, stop referring to *truth.* I shall not any longer ask, 'What is truth?' There are several reasons. My main reason is that I believe that 'What is?' or 'What are?' questions or, in other words, all verbal or definitional questions, should be eliminated. 'What is?' or 'What are?' questions I regard as pseudo-questions; they do not all seem to be so pseudo, but I do think they all are pseudo-questions. Questions such as, 'What is life?' or 'What is matter?' or 'What is mind?' or 'What is logic?' I think should not be asked. They are typically unfruitful questions.

So I think we should also discard the question, 'What is truth?' My first reason (just mentioned) for discarding the question 'What is truth?' one may call 'anti-essentialism'. My second reason is even more important. It is that we should altogether avoid, like the plague, discussing the meaning of words. Discussing the meaning of words is a favorite game of philosophy, past and present: philosophers seem to be addicted to the idea that words and their meaning are important, and are the special concern of philosophy.

I will for your convenience present again here-on the next page-a table which I have used before.

On the left we have *words or concepts and their meanings,* and on the right we have *statements or propositions or theories and their truth.*

IDEAS *that is*	
DESIGNATIONS *or* TERMS *or* CONCEPTS	STATEMENTS *or* PROPOSITIONS *or* THEORIES
may be formulated in	
WORDS	ASSERTIONS
which may be	
MEANINGFUL	TRUE
and their	
MEANING	TRUTH
May be reduced, by way of	
DEFINITIONS	DERIVATIONS
to that of	
UNDEFINED CONCEPTS	PRIMITIVE PROPOSITIONS
The attempt to establish (rather than reduce) by these means their	
MEANING	TRUTH
leads to an infinite regress	

Now I have been taught by the experience of a lifetime in this field that one should always try to get away from the left side of the table and to keep to the right side. One should always keep to assertions, to theories, and the question of their truth. One should never get involved in verbal questions or questions of meaning, and never get interested in words. If challenged by the question of whether a word one uses really means this or perhaps that, then one should say: 'I don't know, and I am not interested in meanings; and if you wish, I will gladly accept your terminology.' This never does any harm. One should never quarrel about words, and never get involved in questions of terminology. One should always keep away from discussing concepts. What we are really interested in, our real problems, are factual problems, or in other words, problems of theories and their truth. We are interested in theories and how they stand up to critical discussion; and our critical discussion is controlled by our interest in truth.

Having said this, I intend now to stop using the word 'truth'.

Our problem is no longer: Is truth correspondence? Is truth coherence? Is truth usefulness? This being so, how can we formulate our real problem?

Our problem can be sharply formulated only by pointing out that the opponents of the correspondence theories all made an *assertion.* They all asserted that there cannot be such a thing as the correspondence between a statement and a fact. This is their central assertion. They say that this concept is meaningless (or that it is undefinable, which, incidentally, in my opinion does not matter, since definitions do not matter) In other words, the whole problem arises because of doubts, or skepticism, concerning correspondence: whether there is such a thing as a correspondence between a statement and a fact. It is quite clear that these doubts are serious (especially in view of the paradox of the liar).

It is also quite clear that, but for these doubts, the upholders of the coherence theory and of the theory of pragmatic usefulness would really have nothing to argue against. Nobody denies that pragmatic usefulness and such matters as predictive power are important. But should there exist something like the *correspondence of a theory to the facts,* then this would obviously be more important than mere self-consistency, and certainly also much more important than coherence with any earlier knowledge' (or 'belief'); for if a theory corresponds to the facts but does not cohere with some earlier knowledge, then this earlier knowledge should be discarded.

Similarly, if there exists something like the correspondence of theory to the facts, then it is clear that a theory which corresponds to the facts will be as a rule very useful; more useful, qua theory, than a theory which does not correspond to the facts. (On the other hand, it may be very useful for a criminal before a court of justice to cling to a theory which does not correspond to the facts; but as it is not *this* kind of usefulness which the pragmatists have in mind, their views raise a question which is very awkward for them: I mean the question, 'Useful for whom?'.)

Although I am an opponent of pragmatism as a philosophy of science, I gladly admit that pragmatism has emphasized something very important: the question whether a theory has some application, whether it has, for example, predictive power. *Praxis, as* I have put it somewhere, is invaluable for the theoretician as a spur and at the same time as a bridle: it is a spur because it suggests new problems to us, and it is a bridle because it may bring us down to earth and to reality if we get lost in over-abstract theoretical flights of our imagination. All this is to be admitted. And yet, it is clear that the pragmatist position will be superseded by a realist position if we can meaningfully say that a statement, or a theory, may or may not correspond to the facts.

Thus the correspondence theory does not deny the importance of the coherence and pragmatist theories, though it does imply that they are not good enough. On the other hand, the coherence and pragmatist theories assert the impossibility or meaninglessness of the correspondence theory.

So without ever mentioning the word 'truth' or asking, 'What does truth mean?' we can see that the central problem of this whole discussion is not the verbal problem of defining 'truth' but the following substantial problem: can there be such a thing as a statement or a theory which corresponds to the facts, or which does not correspond to the facts?

Behind the doubts concerning the possibility of speaking about correspondence, there are various strong arguments.

First of all, there are paradoxes or antinomies which arise out of this correspondence idea. Secondly, there are the countless unsuccessful attempts to say more precisely what the correspondence between a statement and a fact consists of. There is the attempt of Schlick, who said that correspondence is to be explained by a one-one relationship between the linguistic statement and the fact; that is, by uniqueness. A statement, he said, is 'true', or corresponds to the facts, if it stands to the facts of the world in a one-one relationship or in a unique relationship: non-correspondence or 'falsity' is the same as ambiguity. Of course, this is an unacceptable view, for many vague and ambiguous statements (such as 'there are a few people somewhere in America') may correspond to the facts; and vice versa, every general proposition or theory which corresponds to the facts corresponds to many facts, so that there is not a one-one relationship.

Moreover, a statement which does not correspond to the facts may be quite unambiguous. A murderer may say unambiguously, 'I have not killed him.' There is no ambiguity in this assertion; but it does not correspond to the facts. Clearly, Schlick's attempt to explain correspondence misfires. Another even worse attempt is Wittgenstein's. Wittgenstein suggested that a proposition is a picture of reality and that correspondence is a relationship very much like the one that holds between the groove on a gramophone record and the sounds which it denotes: a kind of projective relationship between facts and statements. The untenability of this view can easily be shown. One is reminded of the famous story of Livingstone being introduced by an interpreter to a Negro king whom he asked, 'How are you?' The Negro king answered with one word, and the interpreter began to talk and talk and talk and talk, for ten minutes, translating the word to Livingstone in the form of a long story of the king's sorrows. Then Livingstone asked whether the king was in need of medical assistance, and then the king began to talk and talk and talk and talk and talk. And the interpreter translated it with one word: 'No.'

No doubt this story is invented. But it is well invented; and it illustrates the weakness of the projection theory of language, especially as a theory of the correspondence between a statement and a fact.

But this is not all. The matter is even more serious; namely, Wittgenstein, after having formulated this theory, said that it is impossible to discuss the relationship of language to reality, or to discuss language at all. (Because language cannot be discussed by language.) This is a field in which words fail us. 'It shows itself' is his favorite expression to indicate the failure of words. Any attempt to go deeper into the relationship between language and reality or to discuss language more deeply or statements more deeply is, accordingly, bound to be meaningless. And although he says in the Preface of his book, 'the *truth* of the thoughts that are here set forth seems to me unassailable and definitive', he ends up by saying, and ‘Anybody who understands me eventually recognizes them [the propositions of the Tractatus] as nonsensical (Because talk about language is meaningless.) No doubt this refers, apart from other things, especially to his theory of projection. His remark that his readers will see that what he says is meaningless thus confirms what the opponents of the correspondence theory have always said of the correspondence theory, namely that it is meaningless to speak about the correspondence between a statement and a fact.

So we are back at the real issue. It is this: is there or is there not a tenable correspondence theory? Can we or can we not speak meaningfully of the correspondence between a statement and a fact?

Now my assertion is that Tarski has rehabilitated the correspondence theory. This, I think, is a great achievement, and it is a great philosophical achievement. I say this because it has been denied by many philosophers (for example, by Max Black) that there is something philosophically important in Tarski's achievement.

The key to the rehabilitation of the correspondence theory is a very simple and obvious observation made by Tarski. That is, if I want to speak about correspondence between a statement *S* and a fact *F*, then I have to do so in a language in which I can speak about both: statements such as *S*, and facts such as *F*. This seems to be frightfully trivial; but it is nevertheless decisive. It means that the language in which we speak in explaining correspondence must possess the means needed *to refer* to statements, and to *describe* facts. If I have a language which has both these means at its disposal, so that it can refer to statements and describe facts, then in this language -the metalanguage-I can speak about correspondence between statements and facts without any difficulty, as we shall see.

A metalanguage is a language in which we talk about some other language. For example, a grammar of the German language, written in English, uses English as a metalanguage in order to talk about German. The language about which we talk in the *metalanguage* (in this case English) is usually called *the 'object language'* (in this case German). The characteristic thing about a metalanguage is that it contains (metalinguistic) *names* of words and of statements of the object language, and also (metalinguistic) *predicates,* such as 'noun (of the object language)' or 'verb (of the object language)' or 'statement (of the object language)'. If a metalanguage is to suffice for our purpose it must also, as Tarski points out, contain the usual means necessary to speak about at least all those *facts* about which the object language can speak.

All this is the case if we use English as our metalanguage in order to speak about German (as the object language under investigation).

For example, we shall be able to say in the English metalanguage such things as:

> The German words '*Das Gras ist grün*' form a statement of the German language.
>
> On the other hand, we shall be able to describe in our (English) metalanguage the fact which the German statement '*Das Gras ist grün*' describes. We can describe this fact in English simply by saying that grass is green.
>
> We can now make a statement in the metalanguage about *the correspondence of a statement of the object language to the facts* as follows. We can make the assertion: *The German statement 'Das Gras ist grün' corresponds to the facts if, and only if, grass is green.* (Or: '. . . *only if it is a fact that grass is green.*')

This is very trivial. It is, however, important to realize the following: in our assertion, the words '*Das Gras ist grün*', put within *quotes,* function as a metalinguistic (that is, an *English)* name of a German statement; on the other hand, the English words 'grass is green' occur in our assertion above *without* any quotation marks: they do not function as a name of a statement, but simply as the description of a *fact* (or alleged fact).

This makes it possible for our assertion to express a relationship between a (German) *statement,* and a *fact.* (The *fact* is neither German nor English, although it is, of course, described or spoken about in our metalanguage, which is English: the fact is non-linguistic, it is a fact of the real world, although we need of course a language if we wish to talk about it.) And what our metalinguistic assertion asserts is that a certain (German) statement *corresponds to a certain fact* (a non-linguistic fact, a fact of the real world) under conditions which are precisely stated.

We can, of course, replace the German object language by any other-even by English. Thus we can make the metalinguistic assertion:

> The English statement 'Grass is green' corresponds to the facts if, and only if, grass is green.

This looks even more trivial. But it can hardly be denied; nor can it be denied that it expresses the conditions under which a statement corresponds to the facts.

Generally speaking, let *'S'* be a (metalinguistic) name of a statement of the object language, and let *'f'* be the *abbreviation* of an expression of the metalanguage that describes the (supposed) fact *F* which *S* describes. Then we can make the following metalinguistic assertion:

> A statement *S* of the object language corresponds to the facts if, and only if, *f*. (or: . . . if it is a fact that *f*.)

Note that while '*S*' is here a metalinguistic name of a statement, '*f*' is not a name, but an abbreviation of an expression of the metalanguage describing a certain fact (the fact which we can name '*F*').

We can now say that what Tarski did was to discover that in order to speak about the correspondence between a statement S and a fact *F*, we need a language (a metalanguage) in which we can *speak about* the statement *S and state* the fact *F*. (The former we speak about by using the *name* '*S*', the latter by using a metalinguistic expression '*f*' which *states or describes F*.)

The importance of this discovery is that it dispels all doubt about the meaningfulness of talking about the correspondence of a statement to some fact or facts.

Once this is done, we can; of course, replace the words correspond to the facts' by the words 'is true'.

Tarski, apart from this, introduced a method of giving a *definition* of truth (in the sense of the correspondence theory) for any consistent *formalized system*. But this is not, I think, his main achievement. His main achievement is the rehabilitation of talk about correspondence (and truth). Incidentally, he showed under what circumstances such talk may lead to paradoxes, and how we can avoid these paradoxes; *and he also showed how in ordinary talk about truth we can, and do, avoid paradoxes.*

Once we have settled that we can use 'truth' in the sense of the correspondence of statements to facts, there is really nothing of importance to be added about the word 'truth'. There is no doubt that correspondence to the facts is what we usually call 'truth'; that in ordinary language it is correspondence that we call 'truth', rather than coherence or pragmatic usefulness. A judge who admonishes a witness to speak the truth and nothing but the truth does not admonish the witness to speak what he thinks is useful either for himself or for anybody else. The judge admonishes a witness to speak the truth and nothing but the truth but he does not say, 'All we require of you is that you do not get involved in contradictions', which he would say where he a believer in the coherence theory. But this is not what he demands of the witness.

In other words, the ordinary sense of 'truth' as it is used in courts of law is, no doubt, correspondence. But my main point is that this may be regarded as an afterthought, and as an unimportant afterthought. For if anybody should want to say, 'No, in ordinary language, "truth" is used in a different sense', I should not quarrel with him. I should suggest that we forget all about terminology: I should be prepared to use the terminology of my opponent, pointing out, however, that we have *at least* these three meanings at our disposal: this is the only thing about which I should be prepared to quarrel; but I should refuse to quarrel about words.

I should point out, though, that the correspondence theory of truth is a realistic theory; that is to say, it makes the distinction, which is a realistic distinction, between a theory and the facts which the theory describes; and it makes it possible to say that a theory is true, or false, or that it corresponds to the facts, thus relating the theory to the facts. It allows us to speak of a reality different from the theory. This is the main thing; it is the main point for the realist. The realist wants to have both a theory and the reality or the facts (don't call it 'reality' if you don't like it, just call it 'the facts') which are different from his theory *about* these facts, and which lie can somehow or other compare with the facts, in order to find out whether or not it corresponds to them. Of course, the comparison is always extremely difficult.

One last word about Tarski's theory. Its whole purpose is often misinterpreted: it is wrongly assumed that it is intended to yield a *criterion of truth.* For-coherence was so intended and likewise pragmatic usefulness; they strengthened the traditional view that any serious theory of truth should present us with *a method of deciding* whether or not a given statement is true.

Tarski has proved many things from his definition of truth. Among other things, he has proved that in a sufficiently powerful language (and in every language in which we can formulate mathematical or physical theories) there can be no criterion of truth; that is, no criterion of correspondence: the question of whether a proposition is true is not in general decidable for the languages for which we may form the concept of truth. Thus the concept of truth plays mainly the role of a regulative idea. It helps us in our search for truth that we know there is something like truth or correspondence. It does not give us a means of finding truth, or of being sure that we have found it even if we have found it. So there is no criterion of truth, and we must not ask for a criterion of truth. We must be content with the fact that the idea of truth as correspondence to the facts has been rehabilitated. This has been done by Tarski; and I think that he has thereby rendered an immense service to the realistic outlook.

Although we have no criterion of truth, and no means of being even quite sure of the falsity of a theory, it is easier to find out that a theory is false than to find out that it is true (as I have explained in detail elsewhere). We have even good reasons to think that most of our theories-even our best theories are, strictly speaking, false; for they oversimplify or idealize the facts. Yet a false conjecture may be nearer or less near to the truth. Thus we arrive at the idea of nearness to the truth, or of a better or less good approximation to the truth; that is, at the idea of '*verisimilitude*'. I have tried to show that this idea can be rehabilitated in a way similar to Tarski's rehabilitation of the idea of truth as correspondence to the facts. In order to do so I have used mainly the two Tarskian ideas mentioned here. One is the idea of truth. The other is the idea of logical consequence; or more precisely, of the set of logical consequences of a conjecture, or the content of a conjecture.

By incorporating into logic the idea of verisimilitude or approximation to truth, we make logic even more 'realistic'. For it can now be used to speak about the way in which one theory corresponds better than another to the facts-the facts of the real world.

To sum up. As a realist I look upon logic as the *organon of criticisms* (rather than of proof) in our search for true and highly informative theory as - or at least for new theories that contain more information, and correspond better to the facts, than our older theories. And I look upon criticism, in its turn, as our main instrument in promoting the growth of our knowledge about the world of facts.

Chapter 7

The Analysis of Sensations and the Relation of the Physical to the Psychical

Ernst Mach

I. Introductory Remarks: Anti Metaphysical

1.1.

The great results achieved by physical science in modern times - results not restricted to its own sphere but embracing that of other sciences which employ its help - have brought it about that physical ways of thinking and physical modes of procedure enjoy on all hands unwonted prominence, and that the greatest expectations are associated with their application. In keeping with this drift of modern inquiry, the physiology of the senses, gradually abandoning the method of investigating sensations in themselves followed by men like Goethe, Schopenhauer, and others, but with greatest success by Johannes Muller, has also assumed an almost exclusively physical character. This tendency must appear to us as not altogether appropriate, when we reflect that physics, despite its considerable development, nevertheless constitutes but a portion of a *larger* collective body of knowledge, and that it is unable, with its limited intellectual implements, created for limited and special purposes, to exhaust all the subject-matter in question. Without renouncing the support of physics, it is possible for the physiology of the senses, not only to pursue its own course of development, but also to afford to physical science itself powerful assistance. The following simple considerations will serve to illustrate this relation between the two.

1.2.

Colors, sounds, temperatures, pressures, spaces, times, and so forth, are connected with one another in manifold ways; and with them are associated dispositions of mind, feelings, and volitions. Out of this fabric, that which is relatively more fixed and permanent stands prominently forth, engraves itself on the memory, and expresses itself in language. Relatively greater permanency is exhibited, first, by certain complexes of colors, sounds, pressures, and so forth, functionally connected in time and space, which therefore receive special names, and are called bodies. Absolutely permanent such complexes are not.

My table is now brightly, now dimly lighted. Its temperature varies. It may receive an ink stain. One of its legs may be broken. It may be repaired, polished, and replaced part by part. But, for me, it remains the table at which I daily write.

My friend may put on a different coat. His countenance may assume a serious or a cheerful expression. His complexion, under the effects of light or emotion, may change. His shape may be altered by motion, or be definitely changed. Yet the number of the permanent features presented, compared with the number of the gradual alterations, is always so great, that the latter may be overlooked. It is the same friend with whom I take my daily walk.

My coat may receive a stain, a tear. My very manner of expressing this shows that we are concerned here with a sum-total of permanency, to which the new element is added and from which that which is lacking is subsequently taken away.

Our greater intimacy with this sum-total of permanency, and the preponderance of its importance for me as contrasted with the changeable element, impel us to the partly instinctive, partly voluntary and conscious economy of mental presentation and designation, as expressed in ordinary thought and speech. That which is presented in a single image receives a single designation, a single name.

Further, that complex of memories, moods, and feelings, joined to a particular body (the human body), which is called the "I" or "Ego," manifests itself as relatively permanent. I may be engaged upon this or that subject, I may be quiet and cheerful, excited and ill-humored. Yet, pathological cases apart, enough durable features remain to identify the ego. Of course, the ego also is only of relative permanency.

The apparent permanency of the ego consists chiefly in the single fact of its continuity, in the slowness of its changes. The many thoughts and plans of yesterday that are continued today, and of which our environment in waking hours incessantly reminds us (whence in dreams the ego can be very indistinct, doubled, or entirely wanting), and the little habits that are unconsciously and involuntarily kept up for long periods of time, constitute the groundwork of the ego. There can hardly be greater differences in the egos of different people, than occur in the course of years in one person. When I recall today my early youth, I should take the boy that I then was, with the exception of a few individual features, for a different person, were it not for the existence of the chain of memories. Many an article that I myself penned twenty years ago impresses me now as something quite foreign to myself. The very gradual character of the changes of the body also contributes to the stability of the ego, but in a much less degree than people imagine. Such things are much less analyzed and noticed than the intellectual and the moral ego. Personally, people know themselves very poorly. When I wrote these lines in 1886, Ribot's admirable little book, *The Diseases of Personality* (second edition, Paris, 1888, Chicago, 1895), was unknown to me. Ribot ascribes the principal role in preserving the continuity of the ego to the general sensibility. Generally, I am in perfect accord with his views.

The ego is as little absolutely permanent as are bodies. That which we so much dread in death, the annihilation of our permanency, actually occurs in life in abundant measure. That which is most valued by us, remains preserved in countless copies, or, in cases of exceptional excellence, is even preserved of itself. In the best human being, however, there are individual traits, the loss of which neither he himself nor others need regret. Indeed, at times, death, viewed as liberation from individuality, may even become a pleasant thought. Such reflections of course do not make physiological death any the easier to bear.

After a first survey has been obtained, by the formation of the substance-concepts " body " and " ego " (matter and soul), the will is impelled to a more exact examination of the changes that take place in these relatively permanent existences. The element of change in bodies and the ego, is in fact, exactly what moves the will I to this examination. Here the component parts of the complex are first exhibited as its properties. A fruit is sweet; but it can also be bitter. Also, other fruits may be sweet. The red color we are seeking is found in many bodies. The neighborhood of some bodies is pleasant; that of others, unpleasant. Thus, gradually, different complexes are found to be made up of common elements. The visible, the audible, the tangible, are separated from bodies.

The visible is analyzed into colors and into form. In the manifoldness of the colors, again, though here fewer in number, other component parts are discerned - such as the primary colors, and so forth. The complexes are disintegrated into elements, that is to say, into their ultimate component parts, which hitherto we have been unable to subdivide any further.

The nature of these elements need not be discussed at present; it is possible that future investigations may throw light on it. We need not here be disturbed by the fact that it is easier for the scientist to study relations of relations of these elements than the direct relations between them.

1.3.

The useful habit of designating such relatively permanent compounds by single names, and of apprehending them by single thoughts, without going to the trouble each time of an analysis of their component parts, is apt to come into strange conflict with the tendency to isolate the component parts. The vague image which we have of a given permanent complex, being an image which does not perceptibly change when one or another of the component parts is taken away seems to be something which exists in itself.

Inasmuch as it is possible to take away singly every constituent part without destroying the capacity of the image to stand for the totality and to be recognized again, it is imagined that it is possible to subtract *all* the parts and to have something still remaining. Thus naturally arises the philosophical notion, at first impressive, but subsequently recognized as monstrous, of a “thing-in-itself," different from its "appearance," and unknowable.

Thing, body, matter, are nothing apart from the combinations of the elements, - the colors, sounds, and so forth - nothing apart from their so-called attributes. That protean pseudo-philosophical problem of the single thing with its many attributes arises wholly from a misinterpretation of the fact, that summary comprehension and precise analysis, although both are provisionally justifiable and for many purposes profitable, cannot be carried on simultaneously. A body is one and unchangeable only so long as it is unnecessary to consider its details.

Thus both the earth and a billiard-ball are spheres, if we are willing to neglect all deviations from the spherical form, and if greater precision is not necessary. But when we are obliged to carry on investigations in tomography or microscopy, both bodies cease to be spheres.

1.4.

Man is pre-eminently endowed with the power of voluntarily and consciously determining his own point of view. He can at one time disregard the most salient features of an object, and immediately thereafter give attention to its smallest details; now consider a stationary current, without a thought of its contents (whether heat, electricity or fluidity), and then measure the width of a Fraunhofer line in the spectrum; he can rise at will to the most general abstractions or bury himself in the minutest particulars.

Animals possess this capacity in a far less degree. They do not assume a point of view, but are usually forced to it by their sense-impressions. The baby that does not know its father with his hat on, the dog that is perplexed at the new coat of its master, have both succumbed in this conflict of points of view. Who has not been worsted in similar plights? Even the man of philosophy at times succumbs, as the grotesque problem, above referred to, shows.

In this last case, the circumstances appear to furnish a real ground of justification. Colors, sounds, and the odors of bodies are evanescent. But their tangibility, as a sort of constant nucleus, not readily susceptible of annihilation, remains behind; appearing as the vehicle of the more fugitive properties attached to it. Habit, thus, keeps our thought firmly attached to this central nucleus, even when we have begun to recognize that seeing hearing, smelling, and touching are intimately akin in character.

A further consideration is, that owing to the singularly extensive development of mechanical physics a kind of higher reality is ascribed to the spatial and to the temporal than to colors, sounds, and odors; agreeably to which, the temporal and spatial links of colors, sounds, and odors appear to be more real than the colors, sounds and odors themselves.

The physiology of the senses, however, demonstrates that spaces and times may just as appropriately be called sensations as colors and sounds. But of this later.

1.5.

Not only the relation of bodies to the ego, but the ego itself also, gives rise to similar pseudo - problems, the character of which may be briefly indicated as follows:

> Let us denote the above-mentioned elements by the letters A B C . . ., X L M . . ., a, b, c . . . Let those complexes of colors, sounds, and so forth, commonly called bodies, be denoted, for the sake of clearness, by A B C . .; the complex, known as our own body, which is a part of the former complexes distinguished by certain peculiarities, may be called K L M . . .; the complex composed of volitions, memory-images, and the rest, we shall represent by a b c . . . Usually, now, the complex a , c . . . K L M. . ., as making up the ego, is opposed to the complex A B C . . ., as making up the world of physical objects; sometimes also, a b c . . . is viewed as ego, and K L M . . . A B C . . . as world of physical objects. Now, at first blush, A B C . . . appears independent of the ego, and opposed to it as a separate existence. But this independence is only relative, and gives way upon closer inspection. Much, it is true, may change in the complex a b c . . . without much perceptible change being induced in A B C . . .; and *vice versa*. But many changes in a b c . . . do pass, by way of changes in K L M . . ., to A B C . . .; and *vice versa*. (As, for example, when powerful ideas burst forth into acts, or when our environment induces noticeable changes in our body.) At the same time the group K L M . . . appears to be more intimately connected with a b c . . . and with A B C . . ., than the latter with one another; and their relations find their expression in common thought and speech.

Precisely viewed, however, it appears that the group A B C . . . is always codetermined by K L M. A cube when seen close at hand, looks large; when seen at a distance, small; its appearance to the right eye differs from its appearance to the left; sometimes it appears double; with closed eyes it is invisible. The properties of one and the same body, therefore, appear modified by our own body; they appear conditioned by it. But where, now, is that *same* body, which appears so *different*? All that can be said is, that with different K L M different A B C . . . are associated.

A common and popular way of thinking and speaking is to contrast “appearance “with “reality." A pencil held in front of us in the air is seen by us as straight; dip it into the water, and we see it crooked. In the latter case we say that the pencil *appears* crooked, but is in *reality* straight. But what justifies us in declaring one fact rather than another to be the reality, and degrading the other to the level of appearance? In both cases we have to do with facts which present us with different combinations of the elements, combinations which in the two cases are differently conditioned. Precisely because of its environment the pencil dipped in water is optically crooked; but it is tactually and metrically straight.

An image in a concave or flat mirror is *only* visible, whereas under other and ordinary circumstances a tangible body as well corresponds to the visible image. A bright surface is brighter beside a dark surface than beside one brighter than itself. To be sure, our expectation is deceived when, not paying sufficient attention to the conditions, and substituting for one another different cases of the combination, we fall into the natural error of expecting what we are accustomed to, although the case may be an unusual one.

The facts are not to blame for that. In these cases, to speak of "appearance" may have a practical meaning, but cannot have a scientific meaning. Similarly, the question which is often asked, whether the world is real or whether we merely dream it, is devoid of all scientific meaning. Even the wildest dream is a fact as much as any other. If our dreams were more regular, more connected, more stable, they would also have more practical importance for us. In our waking hours the relations of the elements to one another are immensely amplified in comparison with what they were in our dreams. We recognize the dream for what it is. When the process is reversed, the field of psychic vision is narrowed; the contrast is almost entirely lacking. Where there is no contrast, the distinction between dream and waking, between appearance and reality, is quite otiose and worthless.

The popular notion of an antithesis between appearance and reality has exercised a very powerful influence on scientific and philosophical thought. We see this, for example, in Plato's pregnant and poetical fiction of the Cave, in which, with our backs turned towards the fire, we observe merely the shadows of what passes (*Republic*, vii. 1). But this conception was not thought out to its final consequences, with the result that it has had an unfortunate influence on our ideas about the universe.

The universe, of which nevertheless we are a part, became completely separated from us, and were removed an infinite distance away. Similarly, many a young man, hearing for the first time of the refraction of stellar light, has thought that doubt was cast on the whole of astronomy, whereas nothing is required but an easily effected and unimportant correction to put everything right again.

1.6.

We see an object having a point S. If we touch S, that is, bring it into connexion with our body, we receive a prick. We can see S, without feeling the prick. But as soon as we feel the prick we find S on the skin. The visible point, therefore, is a permanent nucleus, to which the prick is annexed, according to circumstances, as something accidental. From the frequency of analogous occurrences we ultimately accustom ourselves to regard all properties of bodies as "effects "proceeding from permanent nuclei and conveyed to the ego through the medium of the body; which affects we call sensations.

By this operation, however, these nuclei are deprived of their entire sensory content, and converted into mere mental symbols. The assertion, then, is correct that the world consists only of our sensations. In which case we have knowledge *only* of sensations, and the assumption of the nuclei referred to, or of a reciprocal action between them, from which sensations proceed, turns out to be quite idle and superfluous. Such a view can only suit with a half-hearted realism or a half-hearted philosophical criticism.

1.7.

Ordinarily the complex a b c . . . K L M . . . is contrasted as ego with the complex A B C . . . At first only those elements of A B C ... that more strongly alter a b c ... as a prick, a pain, are wont to be thought of as comprised in the ego. Afterwards, however, through observations of the kind just referred to, it appears that the right to annex A B C . . . to the ego nowhere ceases. In conformity with this view the ego can be so extended as ultimately to embrace the entire world. The ego is not sharply marked off, its limits are very indefinite and arbitrarily displaceable Only by failing to observe this fact, and by unconsciously narrowing those limits, while at the same time we enlarge them, arise, in the conflict of points of view, the metaphysical difficulties met with in this connexion.

As soon as we have perceived that the supposed unities "body" and "ego" are only makeshifts, designed for provisional orientation and for definite practical ends (so that we may take hold of bodies, protect ourselves against pain, and so forth), we find ourselves obliged, in many more advanced scientific investigations, to abandon them as insufficient and inappropriate. The antithesis between ego and world, between sensation (appearance) and thing, then vanishes, and we have simply to deal with the connexion of the elements a b c . . . A B C . . . K L M . . ., of which this antithesis was only a partially appropriate and imperfect expression. This connexion is nothing more or less than the combination of the above-mentioned elements with other similar elements (time and space). Science has simply to accept this connexion, and to get its bearings in it, without at once wanting to explain its existence.

On a superficial examination the complex a b c . . . appears to be made up of much more evanescent elements than A B C . . . and K L M . . ., in which last the elements seem to be connected with greater stability and in a more permanent manner (being joined to solid nuclei as it were). Although on closer inspection the elements of all complexes prove to be homogeneous, yet even when this has been recognized, the earlier notion of an antithesis of body and spirit easily slips in again. The philosophical spiritualist is often sensible of the difficulty of imparting the needed solidity to his mind-created world of bodies; the materialist is at a loss when required to endow the world of matter with sensation. The monistic point of view, which reflexion has evolved, is easily clouded by our older and more powerful instinctive notions.

1.8.

The difficulty referred to is particularly felt when we consider the following case. In the complex A B C ... which we have called the world of matter, we find as parts, not only our own body K L M . . ., but also the bodies of other persons (or animals) K' L' M'. . , K" L" M" . ... to which, by analogy, we imagine other a' b' c'..., a" b" c", annexed, similar to a b c . . . So long as we deal with K' L' M' . . ., we find ourselves in a thoroughly familiar province which is at every point accessible to our senses. When, however, we inquire after the sensations or feelings belonging to the body K' L' M' . . ., we no longer find these in the province of sense: we add them in thought. Not only is the domain which we now enter far less familiar to us, but the transition into it is also relatively unsafe. We have the feeling as if we were plunging into an abyss. Persons, who adopt this way of thinking only, will never thoroughly rid themselves of that sense of insecurity, which is a very fertile source of illusory problems.

But we are not restricted to this course. Let us consider, first, the reciprocal relations of the elements of the complex A B C . . ., without regarding K L M . . . (our body). All physical investigations are of this sort. A white ball falls upon a bell; a sound is heard. The ball turns yellow before a sodium lamp, red before a lithium lamp. Here the elements (A B C . . .) appear to be connected only with one another and to be independent of our body (K L M . . .). But if we take santonin, the ball again turns yellow. If we press one eye to the side, we see two balls. If we close our eyes entirely, there is no ball there at all. If we sever the auditory nerve, no sound is heard. The elements =4 B C . . ., therefore, are not only connected with one another, but also with K L M; To this extent, and to this extent *only*, do we call A B C . . . sensations, and regard A B C as belonging to the ego. In what follows, wherever the reader finds the terms "Sensation", "Sensation-complex", used alongside of or instead of the expressions "element", "complex of elements", it must be borne in mind that it is only in the connexion and relation in question, only in their functional dependence, that the elements are sensations. In another functional relation they are at the same time physical objects. We only use the additional term "sensations" to describe the elements, because most people are much more familiar with the elements in question as sensations (colors, sounds, pressures, spaces, times, etc.), while according to the popular conception it is particles of mass that are considered as physical elements, to which the elements, in the sense here used, are attached as "properties" or "effects".

In this way, accordingly, we do not find the gap between bodies and sensations above described, between what is without and what is within, between the material world and the spiritual world. All elements A B C . . ., K L M. ... constitute a *single* coherent mass only, in which, when any one element is disturbed, *all* is put in motion; except that a disturbance in K, L M, . . . has a more extensive and profound action than one in A B C . . . A magnet in our neighborhood disturbs the particles of iron near it; a falling boulder shakes the earth; but the severing of a nerve sets in motion the *whole* system of elements. Quite involuntarily does this relation of things suggest the picture of a viscous mass, at certain places (as in the ego) more firmly coherent than in others? I have often made use of this image in lectures.

1.9.

Thus the great gulf between physical and psychological research persists only when we acquiesce in our habitual stereotyped conceptions. A color is a physical object as soon as we consider its dependence, for instance, upon its luminous source, upon other colors, upon temperatures, upon spaces, and so forth. When we consider, however, its dependence upon the retina (the elements K L M. . .), it is a psychological object, a sensation. Not the subject matter, but the direction of our investigation, is different in the two domains. (Cp. also Chapter II., pp. 43, 44.)

Both in reasoning from the observation of the bodies of other men or animals, to the sensations which they possess, as well as in investigating the influence of our own body upon our own sensations, we have to complete observed facts by analogy. This is accomplished with much greater ease and certainty, when it relates, say, only to nervous processes, which cannot be fully observed in our own bodies - that is, when it is carried out in the more familiar physical domain - than when it is extended to the psychical domain, to the sensations and thoughts of other people. Otherwise there is no essential difference.

1.10.

The considerations just advanced, expressed as they have been in an abstract form, will gain in strength and vividness if we consider the concrete facts from which they flow. Thus, I lie upon my sofa. If I close my right eye, the picture represented in the accompanying cut is presented to my left eye In a frame formed by the ridge of my eyebrow, by my nose, and by my moustache, appears a part of my body, so far as visible, with its environment. My body differs from other human bodies - beyond the fact that every intense motor idea is immediately expressed by a movement of it, and that, if it is touched, more striking changes are determined than if other bodies are touched - by the circumstance, that it is only seen piecemeal, and, especially, is seen without a head. If I observe an element A within my field of vision, and investigate its connexion with another element B within the same field, I step out of the domain of physics into that of physiology or psychology, provided B, to use the opposite expression of a friend of mine made upon seeing this drawing, passes through my skin. Reflexions like that for the field of vision may be made with regard to the province of touch and the perceptual domains of the other senses.

1.11.

Reference has already been made to the different character of the groups of elements denoted by A B C . . . and a b c . . . As a matter of fact, when we see a green tree before us, or remember a green tree, that is, represent a green tree to ourselves, we are perfectly aware of the difference of the two cases. The represented tree has a much less determinate, a much more changeable form; its green is much paler and more evanescent; and, what is of especial note, it plainly appears in a different domain. A movement that we will to execute is never more than a represented movement, and appears in a different domain from that of the executed movement, which always takes place when the image is vivid enough. Now the statement that the elements A and a appear in different domains, means, if we go to the bottom of it, simply this, that these elements are united with different other elements. Thus far, therefore, the fundamental constituents of A B C ... a b c . . . would seem to be *the same* (colors, sounds, spaces, times, motor sensations . . .), and only the character of their connexion different.

Ordinarily pleasure and pain are regarded as different from sensations. Yet not only tactual sensations, but all other kinds of sensations, may pass gradually into pleasure and pain. Pleasure and pain also may be justly termed sensations. Only they are not so well analyzed and so familiar, nor, perhaps, limited to so few organs as the common sensations. In fact, sensations of pleasure and pain, however faint they may be, really constitute an essential part of the content of all so-called emotions. Any additional element that emerges into consciousness when we are under the- influence of emotions may be described as more or less diffused and not sharply localized sensations. William James, and after him Theodore Ribot, have investigated the physiological mechanism of the emotions: they hold that what is essential is purposive tendencies of the body to action - tendencies which correspond to circumstances and are expressed in the organism. Only a part of these emerges into consciousness. We are sad because we shed tears, and not *vice versa*, says James. And Ribot justly observes that a cause of the backward state of our knowledge of the emotions is that we have always confined our observation to so much of these physiological processes as emerges into consciousness. At the same time he goes too far when he maintains that everything psychical is merely "*surajoute*" to the physical, and that it is only the physical that produces effects. For us this distinction is non-existent.

Thus, perceptions, presentations, volitions, and emotions, in short the whole inner and outer world, are put together, in combinations of varying evanescence and permanence, out of a small number of homogeneous elements. Usually, these elements are called sensations. But as vestiges of a one-sided theory inhere in that term, we prefer to speak simply of elements, as we have already done. The aim of all research is to ascertain the mode of connexion of these elements. If it proves impossible to solve the problem by assuming *one* set of such elements, then more than one will have to be assumed. But for the questions under discussion it would be improper to begin by making complicated assumptions in advance.

1.12.

That in this complex of elements, which fundamentally is only one, the boundaries of bodies and of the ego do not admit of being established in a manner definite and sufficient for all cases has already been remarked. To bring together elements that are most intimately connected with pleasure and pain into one ideal mental-economical unity, the ego; this is a task of the highest importance for the intellect working in the service of the pain-avoiding, pleasure-seeking will. The delimitation of the ego, therefore, is instinctively effected, is rendered familiar, and possibly becomes fixed through heredity. Owing to their high practical importance, not only for the individual, but for the entire species, the composites “ego" and "body" instinctively make good their claims, and assert themselves with elementary force. In special cases, however, in which practical ends are not concerned, but where knowledge is an end in itself, the delimitation in question may prove to be insufficient, obstructive, and untenable.

Similarly, class-consciousness, class-prejudice, the feeling of nationality, and even the narrowest-minded local patriotism may have a high importance, for certain purposes. But such attitudes will not be shared by the broad-minded investigator, at least not in moments of research. All such egoistic views are adequate only for practical purposes. Of course, even the investigator may succumb to habit. Trifling pedantries and nonsensical discussions; the cunning appropriation of others' thoughts, with perfidious silence as to the sources; when the word of recognition must be given, the difficulty of swallowing one's defeat, and the too common eagerness at the same time to set the opponent's achievement in a false light: all this abundantly shows that the scientist and scholar have also the battle of existence to fight, that the ways even of science still lead to the mouth, and that the pure impulse towards knowledge is still an ideal in our present social conditions.

The primary fact is not the ego, but the elements (sensations). What was said on p. 21 as to the term “sensation" must be borne in mind? The elements constitute the I. s have the sensation green, signifies that the element green occurs in a given complex of other elements (sensations, memories). When *I* cease to have the sensation green, when *I* die, then the elements no longer occur in the ordinary, familiar association. That is all. Only an ideal mental-economical unity, not a real unity, has ceased to exist. The ego is not a definite, unalterable, sharply bounded unity. None of these attributes are important; for all vary even within the sphere of individual life; in fact their alteration is even sought after by the individual. *Continuity* alone is important. This view accords admirably with the position which Weismann has reached by biological investigations. ("*Zur Frage der Unsterblichkeit der Einzelligen*," *Biolog Centralbl.*, Vol. IV. Nos. 21, 22; compare especially pages 654 and 655, where the scission of the individual into two equal halves is spoken of.) But continuity is only a means of preparing and conserving what is contained in the ego. This content, and not the ego, is the principal thing. This content, however, is not confined to the individual. With the exception of some insignificant and valueless personal memories, it remains presented in others even after the death of the individual. The elements that make up the consciousness of a given individual are firmly connected with one another, but with those of another individual they are only feebly connected, and the connexion is only casually apparent. Contents of consciousness, however, that are of universal significance, break through these limits of the individual, and, attached of course to individuals again, can enjoy a continued existence of an impersonal, superpersonal kind, independently of the personality by means of which they were developed. To contribute to this is the greatest happiness of the artist, the scientist, the inventor, the social reformer, etc.

The ego must be given up. It is partly the perception of this fact, partly the fear of it that has given rise to the many extravagances of pessimism and optimism, and to numerous religious, ascetic, and philosophical absurdities. In the long run we shall not be able to close our eyes to this simple truth, which is the immediate outcome of psychological analysis. We shall then no longer place so high a value upon the ego, which even during the individual life greatly changes, and which, in sleep or during absorption in some idea, just in our very happiest moments, may be partially or wholly absent. We shall then be willing to renounce individual immortality, and not place more value upon the subsidiary elements than upon the principal ones. In this way we shall arrive at a freer and more enlightened view of life, which will preclude the disregard of other egos and the overestimation of our own. The ethical ideal founded on this view of life will be equally far removed from the ideal of the ascetic, which is not biologically tenable for whoever practices it, and vanishes at once with his disappearance, and from the ideal of an overweening

Nietzschean "superman," who cannot, and I hope will not be tolerated by his fellow-men.

If knowledge of the connexion of the elements (sensations) does not suffice us, and we ask, *Who* possesses this connexion of sensations, *Which* experiences it? then we have succumbed to the old habit of subsuming every element (every sensation) under some unanalyzed complex, and we are falling back imperceptibly upon an older, lower, and more limited point of view. It is often pointed out, that a psychical experience which is not the experience of a determinate subject is unthinkable, and it is held that in this way the essential part played by the unity of consciousness has been demonstrated. But the Ego-consciousness can be of many different degrees and composed of a multiplicity of chance memories. One might just as well say that a physical process which does not take place in some environment or other, or at least somewhere in the universe, is unthinkable. In both cases, in order to make a beginning with our investigation, we must be allowed to abstract from the environment, which, as regards its influence, may be very different in different cases, and in special cases may shrink to a minimum. Consider the sensations of the lower animals, to which a subject with definite features can hardly be ascribed. It is out of sensations that the subject is built up, and, once built up, no doubt the subject reacts in turn on the sensations.

The habit of treating the unanalyzed ego complex as an indiscernible unity frequently assumes in science remarkable forms. First, the nervous system is separated from the body as the seat of the sensations. In the nervous system again, the brain is selected as the organ best fitted for this end, and finally, to save the supposed psychical unity, a *point* is sought in the brain as the seat of the soul. But such crude conceptions are hardly fit even to foreshadow the roughest outlines of what future research will do for the connexion of the physical and the psychical. The fact that the different organs and parts of the nervous system are physically connected with, and can be readily excited by, one another, is probably at the bottom of the notion of "psychical unity."

I once heard the question seriously discussed, "How the perception of a large tree could find room in the little head of a man?) Now, although this "problem" is no problem, yet it renders us vividly sensible of the absurdity that can be committed by thinking sensations spatially into the brain. When I speak of the sensations of another person, those sensations are, of course, not exhibited in my optical or physical space; they are mentally added, and I conceive them causally, not spatially, attached to the brain observed, or rather, functionally presented. When I speak of my own sensations, these sensations do not exist spatially in my head, but rather my "head" shares with them the same spatial field, as was explained above. (Compare the remarks on Fig. I on pp. I7-I9 above.)

The unity of consciousness is not an argument in point. Since the apparent antithesis between the real world and the world given through the senses lies entirely in our mode of view, and no actual gulf exists between them, a complicated and variously interconnected content of consciousness is no more difficult to understand than is the complicated interconnection of the world.

If we regard the ego as a real unity, we become involved in the following dilemma: either we must set over against the ego a world of unknowable entities (which would be quite idle and purposeless), or we must regard the whole world, the egos of other people included, as comprised in our own ego (a proposition to which it is difficult to yield serious assent).

But if we take the ego simply as a practical unity, put together for purposes of provisional survey, or as a more strongly cohering group of elements, less strongly connected with other groups of this kind, questions like those above discussed will not arise, and research will have an unobstructed future.

In his philosophical notes Lichtenberg says: "We become conscious of certain presentations that are not dependent upon us; of others that we at least think are dependent upon us. Where is the border-line? We know only the existence of our sensations, presentations, and thoughts. We should say, *It thinks*, just as we say, *It lightens*. It is going too far to say *cogito*, if we translate *cogito* by *I think*. The assumption, or postulation, of the ego is a mere practical necessity." Though the method by which Lichtenberg arrived at this result is somewhat different from ours, we must nevertheless give our full assent to his conclusion.

1.13.

Bodies do not produce sensations, but complexes of elements (complexes of sensations) make up bodies. If, to the physicist, bodies appear the real, abiding existences, whilst the "elements" are regarded merely as their evanescent, transitory appearance, the physicist forgets, in the assumption of such a view, that all bodies are but thought-symbols for complexes of elements (complexes of sensations). Here, too, the elements in question form the real, immediate, and ultimate foundation, which it is the task of physiological-physical research to investigate. By the recognition of this fact, many points of physiology and physics assume more distinct and more economical forms, and many spurious problems are disposed of.

For us, therefore, the world does not consist of mysterious entities, which by their interaction with another, equally mysterious entity, the ego, and produce sensations, which alone are accessible. For us, colors, sounds, spaces, times, are provisionally the ultimate elements, whose given connexion it is our business to investigate.

[I have always felt it as a stroke of special good fortune, that early in life, at about the age of fifteen; I lighted, in the library of my father, on a copy of *Kant's Prolegomena to any Future Metaphysics*. The book made at the time a powerful and ineffaceable impression upon me, the like of which I never afterwards experienced in any of my philosophical reading. Some two or three years later the superfluity of the role played by "the thing in itself" abruptly dawned upon me. On a bright summer day in the open air, the world with my ego suddenly appeared to me as *one* coherent mass of sensations, only more strongly coherent in the ego. Although the actual working out of this thought did not occur until a later period, yet this moment was decisive for my whole view. I had still to struggle long and hard before I was able to retain the new conception in my special subject. With the valuable parts of physical theories we necessarily absorb a good dose of false metaphysics, which it is very difficult to sift out from what deserves to be preserved, especially when those theories have become very familiar to us. At times, too, the traditional, instinctive views would arise with great power and place impediments in my way. Only by alternate studies in physics and in the physiology of the senses, and by historico-physical investigations (since about 1863), and after having endeavored in vain to settle the conflict by a physico-psychological monadology (in my lectures on psycho-physics, in the *Zeitschrift fur praktische Heilkunde*, Vienna, 1863, p. 364), have I attained to any considerable stability in my views. I make no pretensions to the title of philosopher. I only seek to adopt in physics a point of view that need not be changed the moment our glance is carried over into the domain of another science; for, ultimately, all must form one whole. The molecular physics of today certainly does not meet this requirement. What I say I have probably not been the first to say. I also do not wish to offer this exposition of mine as a special achievement. It is rather my belief that everyone will be led to a similar view, which makes a careful survey of any extensive body of knowledge. Avenarius, with whose works I became acquainted in 1883, approaches my point of view (*Philosophie als Denken des Welt nach dem Princip des kleinsten Kraftmasses*, 1876). Also Hering, in his paper on Memory

(*Almanach der Wiener Akademie*, 1870, p. 258; English translation, O. C. Pub. Co., Chicago, 4th edition, enlarged, 1913), and J. Popper in his beautiful book, *Das Rechte zu leben und die Pflicht zu sterben* (Leipzig, 1878, p. 62), have advanced allied thoughts. Compare also my paper *Ueber die okonomische Natur der physikalis der Forschung* (*Almanach der WienerAkadernie*, 1882, p. 179, note; English translation in my *Popular Scientific Lectures*, Chicago, 1894). Finally let me also refer here to the introduction to W. Preyer's *Reine Empfindungslehres* to Riehl's *Freibrurger Antrittsrede*, p. 40, and to R. Wahle's *Gehirn und Bewusstsein*, 1884. My views were indicated briefly in 1872 and 1875, and not expounded at length until 1882 and 1883. I should probably have much additional matter to cite as more or less allied to this line of thought, if my knowledge of the literature were more extensive.]

It is precisely in this that the exploration of reality consists. In this investigation we must not allow ourselves to be impeded by such abridgments and delimitations as body, ego, matter, spirit, etc., which have been formed for special, practical purposes and with wholly provisional and limited, ends in view. On the contrary, the fittest forms of thought must be created in and by that research itself, just as is done in every special science. In place of the traditional, instinctive ways of thought, a freer, fresher view, conforming to developed experience, and reaching out beyond the requirements of practical life, must be substituted throughout.

1.14.

Science always has its origin in the adaptation of thought to some definite field of experience. The results of the adaptation are thought-elements, which are able to represent the whole field. The outcome, of course, is different, according to the character and extent of the field. If the field of experience is enlarged, or if several fields heretofore disconnected are united, the traditional, familiar thought-elements no longer suffice for the extended field. In the struggle of acquired habit with the effort after adaptation, problems arise, which disappear when the adaptation is perfected, to make room for others which have arisen meanwhile.

To the physicist, *qua* physicist, the idea of "body" is productive of a real facilitation of view, and is not the cause of disturbance. So, the person with purely practical aims, is materially supported also by the idea of the *I* or ego. For, unquestionably, every form of thought that has been designedly or undesignedly constructed for a given purpose possesses for that purpose a *permanent* value. When, however, physics and psychology meet, the ideas held in the one domain prove to be untenable in the other. From the attempt at mutual adaptation arise the various atomic and monadistic theories - which, however, never attain their end. If we regard sensations, in the sense above defined (p. 13), as the elements of the world, the problems referred to appear to be disposed of in all essentials, and the first and most important adaptation to be consequently effected. This fundamental view (without any pretension to being a philosophy for all eternity) can at present be adhered to in all fields of experience; it is consequently the one that accommodates itself with the least expenditure of energy, that is, more economically than any other, to the present temporary collective state of knowledge Furthermore, in the consciousness of its purely economical function, this fundamental view is eminently tolerant. It does not obtrude itself into fields in which the current conceptions are still adequate. It is also ever ready, upon subsequent extensions of the field of experience, to give way before a better conception.

The presentations and conceptions of the average man of the world are formed and dominated, not by the full and pure desire for knowledge as an end in itself, but by the struggle to adapt himself favorably to the conditions of life. Consequently they are less exact, but at the same time also they are preserved from the monstrosities which easily result from a one-sided and impassioned pursuit of a scientific or philosophical point of view. The unprejudiced man of normal psychological development takes the elements which we have called A B C . . . to be spatially contiguous and external to the elements K L M . . . and he holds this view *immediately*, and not by any process of psychological projection or logical inference or construction; even were such a process to exist, he would certainly not be conscious of it. He sees, then, an "external world" A B C . . . different from his body K L M. and existing outside it. As he does not observe at first the dependence of the A B C's . . . on the K L M's . . . (which are always repeating themselves in the same way and consequently receive little attention), but is always dwelling upon the fixed connexion of the A B C's . . . with one another, there appears to him a world of things independent of his Ego. This Ego is formed by the observation of the special properties of the particular thing K L M . . . with which pain, pleasure, feeling, will, etc., are intimately connected. Further, he notices things K' L' M', K" L" M", which behave in a manner perfectly analogous to K L M, and whose behavior he thoroughly understands as soon as he has thought of analogous feelings, sensations, etc., as attached to them in the same way as he observed these feelings, sensations, etc., to be attached to himself. The analogy impelling him to this result is the same as determines him, when he has observed that a wire possesses *all* the properties of a conductor charged with an electric current, except one which has not yet been directly demonstrated, to conclude that the wire possesses this one property as well. Thus, since he does not perceive the sensations of his fellowmen or of animals but only supplies them by analogy, while he infers from the behavior of his fellow-men that they are in the same position over against himself, he is led to ascribe to the sensations, memories, etc., a particular A B C . . . K L M . . . of a different nature, always differently conceived according to the degree of civilization he has reached; but this process, as was shown above, is unnecessary, and in science leads into a maze of error, although the falsification is of small significance for practical life.

These factors, determining as they do the intellectual outlook of the plain man, make their appearance alternately in him according to the requirements of practical life for the time being, and persist in a state of nearly stable equilibrium. The scientific conception of the world, however, puts the emphasis now upon one, now upon the other factor, makes sometimes one and sometimes the other its starting-point, and, in its struggle for greater precision, unity and consistency, tries, so far as seems possible, to thrust into the background all but the most indispensable conceptions. In this way dualistic and monistic systems arise.

The plain man is familiar with blindness and deafness, and knows from his everyday experience that the look of things is influenced by his senses; but it never occurs to him to regard the whole world as the creation of his senses. He would find an idealistic system, or such a monstrosity as solipsism, intolerable in practice.

It may easily become a disturbing element in unprejudiced scientific theorizing when a conception which is adapted to a particular and strictly limited purpose is promoted in advance to be the foundation of *all* investigation. This happens, for example, when all experiences are regarded as "effects" of an external world extending into consciousness. This conception gives us a tangle of metaphysical difficulties which it seems impossible to unravel. But the specter vanishes at once when we look at the matter as it were in a mathematical light, and make it clear to ourselves that all that is valuable to us is the discovery of *functional relations*, and that what we want to know is merely the dependence of experiences or one another. It then becomes obvious that the reference to unknown fundamental variables which are not given (things-in-themselves) is purely fictitious and superfluous. But even when we allow this fiction, uneconomical though it be, to stand at first, we can still easily distinguish different classes of the mutual dependence of the elements of "the facts of consciousness"; and this alone is important for us.

A B C . . . K L M a b c . . .
K' L' M' ... a' b' c'...
K" L" M"... a" B" C"

The system of the elements is indicated in the above scheme. Within the space surrounded by a single line lie the elements which belong to the sensible world, - the elements whose regular connexion and peculiar dependence on one another represent both physical (lifeless) bodies and the bodies of men, animals and plants. All these elements, again, stand in a relation of quite peculiar dependence to certain of the elements K L M - the nerves of our body, namely - by which the facts of sense-physiology are expressed. The space surrounded by a double line contains the elements belonging to the higher psychic life, memory-images and presentations, including those which we form of the psychic life of our fellow-men. These may be distinguished by accents. These presentations, again, are connected with one another in a different way (association, fancy) from the sensational elements A B C . . . K L M; but it cannot be doubted that they are very closely allied to the latter, and that in the last resort their behavior is determined by A B C . . . K L M (the totality of the physical world), and especially by our body and nervous system. The presentations a' b' c' of the contents of the consciousness of our fellow-men play for us the part of intermediate substitutions, by means of which the behavior of our fellow-men, - the functional relation of K' L' M' to A B C - becomes intelligible, in so far as in and for itself(physically) it would remain unexplained.

It is therefore important for us to recognize that in all questions in this connexion, which can be intelligibly asked and which can interest us, everything turns on taking into consideration different *ultimate variables* and different *relations of dependence*. That is the main point. Nothing will be changed in the actual facts or in the functional relations, whether we regard all the data as contents of consciousness or as partially so, or as completely physical.

The biological task of science is to provide the fully developed human individual with as perfect a means of orientating himself as possible. No other scientific ideal can be realized, and any other must be meaningless.

The philosophical point of view of the average man - if that term may be applied to his naive realism - has a claim to the highest consideration. It has arisen in the process of immeasurable time without the intentional assistance of man. It is a product of nature, and is preserved by nature. Everything at philosophy has accomplished - though we may admit the biological justification of every advance, nay, of every error - is, as compared with it, but an insignificant and ephemeral product of art. The fact is, every thinker, every philosopher, the moment he is forced to abandon his one-sided intellectual occupation by practical necessity, immediately returns to the general point of view of mankind. Professor X., who theoretically believes himself to be a solipsist, is certainly not one in practice when he has to thank a Minister of State for a decoration conferred upon him, or when he lectures to an audience. The Pyrrhonist who is cudgeled in Moliere's *Le Mariage force*, does not go on saying "*Il me semble que vous me battez*," but takes his beating as really received.

Nor is it the purpose of these “introductory remarks” to discredit the standpoint of the plain man. The task which we have set ourselves is simply to show why and for what purpose we hold that standpoint during most of our lives, and why and for what purpose we are provisionally obliged to abandon it. No point of view has absolute, permanent validity. Each has importance only for some given end ...

Chapter 8

OUTLINE OF AN ANARCHISTIC THEORY OF KNOWLEDGE

Paul Feyerabend

- Science is an essentially anarchistic enterprise: theoretical anarchism is more humanitarian and more likely to encourage progress than its law-and-order alternatives.
- This is shown both by an examination of historical episodes and by an abstract analysis of the relation between idea and action. The only principle that does not inhibit progress is: *anything goes.*
- For example, we may use hypotheses that contradict well-confirmed theories and/or well-established experimental results. We may advance science by proceeding counter-inductively.
- The consistency condition which demands that new hypotheses agree with accepted *theories* is unreasonable because it preserves the older theory, and not the better theory. Hypotheses contradicting well-confirmed theories give us evidence that cannot be obtained in any other way. Proliferation of theories is beneficial for science, while uniformity impairs its critical power. Uniformity also endangers the free development of the individual.
- There is no idea, however ancient and absurd, that is not capable of improving our knowledge. The whole history of thought is absorbed into science and is used for improving every single theory. Nor is political interference rejected. It may be needed to overcome the chauvinism of science that resists alternatives to the status quo.

- No theory ever agrees with all the *facts* in its domain, yet it is not always the theory that is to blame. Facts are constituted by older ideologies, and a clash between facts and theories may be proof of progress. It is also a first step in our attempts to find the principles implicit in familiar observational notions.
- As an example of such an attempt I examine the *tower argument* which the Aristotelians used to refute the motion of the earth. The argument involves *natural interpretations* - ideas so closely connected with observations that it needs a special effort to realize their existence and to determine their content. Galileo identifies the natural interpretations which are inconsistent with Copernicus and replaces them by others.
- The new natural interpretations constitute a new and highly abstract observation language. They are introduced *and concealed* so that one falls to notice the change that has taken place (method of anamnesis). They contain the idea of the *relativity of all motion* and the *law of circular inertia.*
- Initial difficulties caused by the change are defused by *ad hoc hypotheses,* which thus turn out occasionally to have a positive function; they give new theories a breathing space, and they indicate the direction of future research.
- In addition to natural interpretations, Galileo also changes *sensations* that seem to endanger Copernicus. He admits that there are such sensations, he praises Copernicus for having disregarded them, and he claims to have removed them with the help of the *telescope.* However, he offers no *theoretical* reasons why the telescope should be expected to give a true picture of the sky.
- Nor does the initial *experience* with the telescope provide such reasons. The first telescopic observations of the sky are indistinct, indeterminate, and contradictory and in conflict with what everyone can see with his unaided eyes. And, the only theory that could have helped to separate telescopic illusions from veridical phenomena was refuted by simple tests.
- On the other hand, there are some telescopic phenomena which are plainly Copernican. Galileo introduces these phenomena as independent evidence for Copernicus while the situation is rather that one refuted view - Copernicanism - has a certain similarity with phenomena emerging from another refuted view - the idea that

telescopic phenomena are faithful images of the sky. Galileo prevails because of his style and his clever techniques of persuasion, because he writes in Italian rather than in Latin, and because he appeals to people who are temperamentally opposed to the old ideas and the standards of learning connected with them.

- Such 'irrational' methods of support are needed because of the 'uneven development' (Marx, Lenin) of different parts of science. Copernicanism and other essential ingredients of modern science survived only because reason was frequently overruled in their past.
- Galileo's method works in other fields as well. For example, it can be used to eliminate the existing arguments against materialism, and to put an end to the *philosophical* mind/body problem (the corresponding *scientific* problems remain untouched, however).
- The results obtained so far suggest abolishing the distinction between a context of discovery and a context of justification and disregarding the related distinction between observational terms and theoretical terms. Neither distinction plays a role in scientific practice. Attempts to enforce them would have disastrous consequences.
- Finally, the discussion in Chapters 6-13 shows that Popper's version of Mill's pluralism is not in agreement with scientific practice and would destroy science as we know it. Given science, reason cannot be universal and unreason cannot be excluded. This feature of science calls for an anarchistic epistemology. The realization that science is not sacrosanct, and that the debate between science and myth has ceased without having been won by either side, further strengthens the case for anarchism.
- Even the ingenious attempt of Lakatos to construct a methodology that (a) does not issue orders and yet (b) puts restrictions upon our knowledge-increasing activities, does not escape this conclusion. For Lakatos' philosophy appears liberal only because it is an *anarchism in disguise.* And his standards which are abstracted from modern science cannot be regarded as neutral arbiters in the issue between modern science and Aristotelian science, myth, magic, religion, etc.
- Moreover, these standards, which involve a comparison of content classes, are not always *applicable.* The content classes of certain theories are incomparable in the sense that none of the usual logical relations (inclusion, exclusion, overlap) can be said to hold between them. This occurs when we compare myths with science. It also

occurs in the most advanced, most general and therefore most mythological parts of science itself.

- Thus science is much closer to myth than a scientific philosophy is prepared to admit. It is one of the many forms of thought that have been developed by man, and not necessarily the best. It is conspicuous, noisy, and impudent, but it is inherently superior only for those who have already decided in favor of a certain ideology, or who have accepted it without having ever examined its advantages and its limits. And as the accepting and rejecting of ideologies should be left to the individual it follows that the separation of state and *church* must be supplemented by the separation of state and *science,* that most recent, most aggressive, and most dogmatic religious institution. Such a separation may be our only chance to achieve a humanity we are capable of, but have never fully realized.

The idea that science can, and should, be run according to fixed and universal rules, is both unrealistic and pernicious. It is *unrealistic,* for it takes too simple a view of the talents of man and of the circumstances which encourage, or cause, their development. And it is *pernicious*, for the attempt to enforce the rules is bound to increase our professional qualifications at the expense of our humanity. In addition, the idea is *detrimental to science,* for it neglects the complex physical and historical conditions which influence scientific change. It makes our science less adaptable and more dogmatic: every methodological rule is associated with cosmological assumptions, so that using the rule we take it for granted that the assumptions are correct. Naive falsificationism takes it for granted that the laws of nature are manifest and not hidden beneath disturbances of considerable magnitude. Empiricism takes it for -ranted that sense experience is a better mirror of the world than pure thought. Praise of argument takes it for granted that the artifices of Reason give better results than the unchecked play of our emotions. Such assumptions may be perfectly plausible *and even true.* Still, one should occasionally put them to a test. Putting them to a test means that we stop using the methodology associated with them, start doing science in a different way and see what happens. Case studies such as those reported in the preceding chapters show that such tests occur all the time, and that they speak *against* the universal validity of any rule. All methodologies have their limitations and the only 'rule' that survives is 'anything goes'.

The change of perspective brought about by these discoveries leads once more to the long-forgotten problem of the excellence of science. It leads to it for the first time in *modern* history, for modern science *overpowered* its opponents, it did not *convince* them. Science took over by *force,* not by argument (this is especially true of the former colonies where science and the religion of brotherly love were introduced as a matter of course and without consulting, or arguing with, the inhabitants). Today we realize that rationalism, being bound to science, cannot give us any assistance in the issue between science and myth and we also know, from inquiries of an entirely different kind, that myths are vastly better than rationalists have dared to admit.' Thus we are now *forced* to raise the question of the excellence of science. An examination then reveals that science and myth overlap in many ways, that the differences we think we perceive are often local phenomena which may turn into similarities elsewhere and that fundamental discrepancies are results of different *aims* rather than of different methods trying to reach one and the same 'rational' end (such as, for example, 'progress', or increase of content, or 'growth').

To show the surprising similarities of myth and science, I shall briefly discuss an interesting paper by Robin Horton, entitled 'African Traditional Thought and Western Science'. Horton examines African mythology and discovers the following features: the quest for theory is a quest for unity underlying apparent complexity. The theory places things in a causal context that is wider than the causal context provided by common sense: both science and myth cap common sense with a theoretical superstructure. There are theories of different degrees of abstraction and they are used in accordance with the different requirements of explanation that arise. Theory construction consists in breaking up objects of common sense and in reuniting the elements in a different way. Theoretical models start from analogy but they gradually move away from the pattern on which the analogy was based. And so on.

These features, which emerge from case studies no less careful and detailed than those of Lakatos, refute the assumption that science and myth obey different principles of formation (Cassirer), that myth proceeds without reflection (Dardel), or speculation (Frankfort, occasionally). Nor can we accept the idea, found in Malinowski but also in classical scholars such as Harrison and Cornford that myth has an essentially pragmatic function or is based on ritual. Myth is much closer to science than one would expect from a philosophical discussion. It is closer to science than even Horton himself is prepared to admit.

To see this, consider some of the *differences* Horton emphasizes. According to Horton, the central ideas of a myth are regarded as sacred. There is anxiety about threats to them. One 'almost never finds a confession of ignorance and events 'which seriously defy the established lines of classification in the culture where they occur' evoke a 'taboo reaction'. Basic beliefs are protected by this reaction as well as by the device of 'secondary elaborations' which, in our terms, arc series of *ad hoc* hypotheses. Science, on the other hand, is characterized by an essential skepticism; 'when failures start to come thick and fast, defense of the theory switches inexorably to attack on it'.' This is possible because of the 'openness' of the scientific enterprise, because of the pluralism of ideas it contains and also because whatever defies or fails to fit into the established category system is not something horrifying, to be isolated or expelled. On the contrary, it is an intriguing 'phenomenon' - a starting-point and a challenge for the invention of new classifications and new theories. We can see that Horton has read his Popper well. A field study of science itself shows a very different picture.

Such a study reveals that, while some scientists may proceed as described, the great majority follow a different path. Skepticism is at a minimum; it is directed against the view of the opposition and against minor ramifications of one's own basic ideas, never against the basic ideas themselves. Attacking the basic ideas evokes taboo reactions which are no weaker than are the taboo reactions in so-called "primitive societies." Basic beliefs are protected by this reaction as well as by secondary elaborations, as we have seen, and whatever fails to fit into the established category system or is said to be incompatible with this system is either viewed as something quite horrifying or, more frequently, it *is simply declared to be non-existent.* Nor is science prepared to make 'a theoretical pluralism the foundation of research. Newton reigned for more than 150 years; Einstein briefly introduced a more liberal point of view only to be succeeded by the Copenhagen Interpretation. The similarities between science and myth are indeed astonishing.

But the fields are even more closely related. The massive dogmatism I have described is not just *a fact;* it has also a most important function. *Science would be impossible without it.* 'Primitive' thinkers showed greater insight into the nature of knowledge than their 'enlightened' philosophical rivals. It is, therefore, necessary to re-examine our attitude towards myth, religion, magic, witchcraft and towards all those ideas which rationalists would like to see forever removed from the surface of the earth (without having so much as looked at them - a typical taboo reaction).

There is another reason why such a re-examination is urgently required. The rise of modern science coincides with the suppression of non-Western tribes by Western invaders. The tribes are not only physically suppressed; they also lose their intellectual independence and are forced to adopt the bloodthirsty religion of brotherly love - Christianity. The most intelligent members get an extra bonus: they are introduced into the mysteries of Western Rationalism and its peak - Western Science. Occasionally this leads to an almost unbearable tension with tradition (Haiti). In most cases the tradition disappears without the trace of an argument, one simply becomes a slave both in body and in mind. Today this development is gradually reversed - with great reluctance, to be sure, but it is reversed. Freedom is regained, old traditions are rediscovered, both among the minorities in Western countries and among large populations in non-Western continents. *But science still reigns supreme.* It reigns supreme because its practitioners are *unable to understand, and unwilling to condone,* different ideologies, because they have *the power* to enforce their wishes, and because they *use* this power ' just as their ancestors used *their* power to force Christianity on the peoples they encountered during their conquests. Thus, while an American can now choose the religion he likes, he is still not permitted to demand that his children learn magic rather than science at school. There is a separation between state and church; there is no separation between state and science.

And yet science has no greater authority than any other form of life. Its aims are certainly not more important than are the aims that guide the lives in a religious community or in a tribe that is united by a myth. At any rate, they have no business restricting the lives, the thoughts, and the education of the members of a free society where everyone should have a chance to make up his own mind and to live in accordance with the social beliefs he finds most acceptable. The separation between state and church must therefore be complemented by the separation between state and science.

We need not fear that such a separation will lead to a breakdown of technology. There will always be people who prefer being scientists to being the masters of their fate and who gladly submit to the meanest kind of (intellectual and institutional) slavery provided they are paid well and provided also there are some people around who examine their work and sing their praise. Greece developed and progressed because it could rely on the services of unwilling slaves. We shall develop and progress with the help of the numerous willing slaves in universities and laboratories who provide us with pills, gas, electricity, atom bombs, frozen dinners and, occasionally, with a few interesting fairy-tales. We shall treat these slaves well, we shall even listen to them, for they have occasionally some interesting stories to tell, but we shall *not* permit them to impose their ideology on our children in the guise of 'progressive' theories of education. We shall not permit them to teach the fancies of science as if they were the only factual statements in existence. This separation of science and state may be our only chance to overcome the hectic barbarism of our scientific-technical age and to achieve a humanity we are capable of, but have never fully realized. Let us, therefore, in conclusion review the arguments that can be adduced for such a procedure.

The image of 20th-century science in the minds of scientists and laymen is determined by technological miracles such as color television, the moon shots, the infra-red oven, as well as by a somewhat vague but still quite influential rumors, or fairy-tale, concerning the manner in which these miracles are produced.

According to the fairy-tale the success of science is the result of a subtle, but carefully balanced combination of inventiveness and control. Scientists have *ideas.* And they have special *methods* for improving ideas. The theories of science have passed the test of method. They give a better account of the world than ideas which have not passed the test.

The fairy-tale explains why modern society treats science in a special way and why it grants it privileges not enjoyed by other institutions.

Ideally, the modern state is ideologically neutral. Religion, myth, prejudices do have an influence, but only in a roundabout way, through the medium of politically influential *parties.* Ideological principles *may* enter the governmental structure, but only via a majority vote, and after a lengthy discussion of possible consequences. In our schools the main religions are taught as *historical phenomena.* They are taught as parts of the truth only if the parents insist on a more direct mode of instruction. It is up to them to decide about the religious education of their children. The financial support of ideologies does not exceed the financial support granted to parties and to private groups. State and ideology, state and church, state and myth, are carefully separated.

State and science, however, work closely to-ether. Immense sums are spent on the improvement of scientific ideas. Bastard subjects such as the philosophy of science which have not a single discovery to their credit profit from the boom of the sciences. Even human relations are dealt with in a scientific manner, as is shown by education programmes, proposals for prison reform, army training, and so on. Almost all scientific subjects are compulsory subjects in our schools. While the parents of a six-year-old child can decide to have him instructed in the rudiments of Protestantism, or in the rudiments of the Jewish faith, or to omit religious instruction altogether, they do not have a similar freedom in the case of the sciences. Physics, astronomy, history *must* be learned. They cannot be replaced by magic, astrology, or by a study of legends.

Nor is one content with a merely *historical* presentation of physical (astronomical, historical, etc.) facts and principles. One does not say: *some people believe* that the earth moves round the sun while others regard the earth as a hollow sphere that contains the sun, the planets, the fixed stars. One says: the earth *moves* round the sun - everything else is sheer idiocy.

Finally, the manner in which we accept or reject scientific ideas is radically different from democratic decision procedures. We accept scientific laws and scientific facts; we teach them in our schools, we make them the basis of important political decisions, but without ever having subjected them to a vote. *Scientists* do not subject them to a vote - or at least this is what they say - *and laymen* certainly do not subject them to a vote. Concrete proposals are occasionally discussed, and a vote is suggested. But the procedure is not extended to general theories and scientific facts. Modern society is 'Copernican' not because Copernicanism has been put on a ballot, subjected to a democratic debate and then voted in with a simple majority; it is 'Copernican' because the *scientists* are Copernicans and because one accepts their cosmology as uncritically as one once accepted the cosmology of bishops and cardinals.

Even bold and revolutionary thinkers bow to the judgment of science. Kropotkin wants to break up all existing institutions - but he does not touch science. Ibsen goes very far in unmasking the conditions of contemporary humanity - but he still retains science as a measure of the truth. Evans-Pritchard, Lévi-Strauss and others have recognized that 'Western Thought', far from being a lonely peak of human development, is troubled by problems not found in other ideologies - but they exclude science from their relativization of all forms of thought. Even for them science is a *neutral structure containing positive knowledge* that is independent of culture, ideology, prejudice.

The reason for this special treatment of science is, of course, our little fairy-tale: if science has found a method that turns ideologically contaminated ideas into true and useful theories, then it is indeed not mere ideology, but an objective measure of all ideologies. It is then not subjected to the demand for a separation between state and ideology.

But the fairy-tale is false, as we have seen. There is no special method that guarantees success or makes it probable. Scientists do not solve problems because they possess a magic wand - methodology, or a theory of rationality - but because they have studied a problem for a long time, because they know the situation fairly well, because they are not too dumb (though that is rather doubtful nowadays when almost anyone can become a scientist), and because the excesses of one scientific school are almost always balanced by the excesses of some other school. (Besides, scientists only rarely solve their problems, they make lots of mistakes, and many of their solutions are quite useless.) Basically there is hardly any difference between the process that leads to the announcement of a new scientific law and the process preceding passage of a new law in society: one informs either all citizens or those immediately concerned, one collects 'facts' and prejudices, one discusses the matter, and one finally votes. But while a democracy makes some effort to *explain* the process so that everyone can understand it, scientists either *conceal* it, or *bend* it, to make it fit their sectarian interests.

No scientist will admit that voting plays a role in his subject. Facts, logic, and methodology alone decide - this is what the fairy-tale tells us. But how do facts decide? What is their function in the advancement of knowledge? We cannot *derive* our theories from them. We cannot give a *negative* criterion by saying, for example, that good theories are theories which can be refuted, but which are not yet contradicted by any fact. A principle of falsification that removes theories because they do not fit the facts would have to remove the whole of science (or it would have to admit that large parts of science are irrefutable). The hint that a good theory *explains more* than its rivals is not very realistic either. True: new theories often predict new things - but almost always at the expense of things already known. Turning to logic we realize that even the simplest demands *are not* satisfied in scientific practice, and *could not be* satisfied, because of the complexity of the material. The ideas which scientists use to present the known and to advance into the unknown are only rarely in agreement with the strict injunctions of logic or pure mathematics and the attempt to make them conform would rob science of the elasticity without which progress cannot be achieved. We see: facts alone are not strong enough for making us accept, or reject, scientific theories, the range they leave to thought is *too wide;* logic and methodology eliminate too much, they are *too narrow.* In between these two extremes lies the ever-changing domain of human ideas and wishes. And a more detailed analysis of successful moves in the game of science ('successful' from the point of view of the scientists themselves) shows indeed that there is a wide range of freedom that *demands* a multiplicity of ideas and *permits* the application of democratic procedures (ballot-discussion-vote) but that is actually closed by power politics and propaganda. *This is where the fairy-tale of a special method assumes its decisive function.* It conceals the freedom of decision which creative scientists and the general public have even inside the most rigid and the most advanced parts of science by a recitation of 'objective' criteria and it thus protects the big-shots (Nobel Prize winners; heads of laboratories, of organizations such as the AMA, of special schools; 'educators'; etc.) from the masses (laymen; experts in non-scientific fields; experts in other fields of science): only those citizens count who were subjected to the pressures of scientific institutions (they have undergone a long process of education), who succumbed to these pressures (they have passed their examinations), and who are now firmly convinced of the truth of the fairy-tale. This is how scientists have deceived themselves and everyone else about their business, but without any real disadvantage: they have more money, more authority, more sex appeal than they deserve, and the most stupid procedures and the most laughable results in their domain are surrounded with an aura of excellence. It is time to cut them down in size, and to give them a more modest position in society.

This advice, which only few of our well-conditioned contemporaries are prepared to accept, seems to clash with certain simple and widely-known facts.

Is it not a fact that a learned physician is better equipped to diagnose and to cure an illness than a layman or the medicine-man of a primitive society? Is it not a fact that epidemics and dangerous individual diseases have disappeared only with the beginning of modern medicine? Must we not admit that technology has made tremendous advances since the rise of modern science? And are not the moon-shots a most impressive and undeniable proof of its excellence? These are some of the questions which are thrown at the impudent wretch who dares to criticize the special position of the sciences.

The questions reach their polemical aim only if one assumes that the results of science *which no one will deny* have arisen without any help from non-scientific elements, and that they cannot be improved by an admixture of such elements either. 'Unscientific' procedures such as the herbal lore of witches and cunning men, the astronomy of mystics, and the treatment of the ill in primitive societies are totally without merit. *Science alone* gives us a useful astronomy, an effective medicine, a trustworthy technology. One must also assume that science owes its success to the correct method and not merely to a lucky accident. It was not a fortunate cosmological guess that led to progress, but the correct *and cosmologically neutral* handling of data. These are the assumptions we must make to give the questions the polemical force they are supposed to have. Not a single one of them stands up to closer examination.

Modern astronomy started with the attempt of Copernicus to adapt the old ideas of Philolaus to the needs of astronomical predictions. Philolaus was not a precise scientist, he was a muddle-headed Pythagorean, as we have seen, and the consequences of his doctrine were called 'incredibly ridiculous' by a professional astronomer such as Ptolemy. Even Galileo, who had the much improved Copernican version of Philolaus before him, says: 'There is no limit to my astonishment when I reflect that Aristarchus and Copernicus were able to make reason to conquer sense that, in defiance of the latter, the former became mistress of their belief' *(Dialogue,* 328). 'Sense' here refers to the experiences which Aristotle and others had used to show that the earth must be at rest. The 'reason' which Copernicus opposes to their arguments is the very mystical reason of Philolaus combined with an equally mystical faith ('mystical' from the point of view of today's rationalists) in the fundamental character of circular motion. I have shown that modern astronomy and modern dynamics could not have advanced without this unscientific use of antediluvian ideas.

While astronomy profited from Pythagoreanism and from the Platonic love for circles, medicine profited from herbalism, from the psychology, the metaphysics, the physiology of witches, midwives, cunning men, wandering druggists. It is well known that 16th- and 17th-century medicine while theoretically hypertrophic was quite helpless in the face of disease (and stayed that way for a long time after the 'scientific revolution'). Innovators such as Paracelsus fell back on the earlier ideas and improved medicine. Everywhere science is enriched by unscientific methods and unscientific results, while procedures which have often been regarded as essential parts of science are quietly suspended or circumvented.

The process is not restricted to the early history of modern science. It is not merely a consequence of the primitive state of the sciences of the 16th and 17th centuries. Even today science can and does profit from an admixture of unscientific ingredients. An example which was discussed above, in Chapter 4, is the revival of traditional medicine in Communist China. When the Communists in the fifties forced hospitals and medical schools to teach the ideas and the methods contained in the *Yellow Emperor's Textbook of Internal Medicine* and to use them in the treatment of patients, many Western experts (among them Eccles, one of the 'Popperian Knights') were aghast and predicted the downfall of Chinese medicine. What happened was the exact opposite. Acupuncture, moxibustion, pulse diagnosis have led to new insights, new methods of treatment, new problems both for the Western and for the Chinese physician.

And those who do not like to see the state meddling in scientific matters should remember the sizeable chauvinism of science: for most scientists the slogan 'freedom for science' means the freedom to indoctrinate not only those who have joined them, but the rest of society as well. Of course - not every mixture of scientific and non-scientific elements is successful (example: Lysenko). But science is not always successful either. If mixtures are to be avoided because they occasionally misfire, then pure science (if there is such a thing) must be avoided as well. (It is not the *interference* of the state that is objectionable in the Lysenko case, but the *totalitarian* interference that kills the opponent instead of letting him go his own way.)

Combining this observation with the insight that science has no special method, we arrive at the result that the separation of science and non-science is not only artificial but also detrimental to the advancement of knowledge. If we want to understand nature, if we want to master our physical surroundings, then we must use *all ideas,* all methods, and not 'just a small selection of them. The assertion, however, that there is no knowledge outside science - *extra scientiam nulla salus* - is nothing but another and most convenient fairy-tale. Primitive tribes have more detailed classifications of animals and plants than contemporary scientific zoology and botany, they know remedies whose effectiveness astounds physicians (while the pharmaceutical industry already smells here a new source of income), they have means of influencing their fellow men which science for a long time regarded as non-existent (Voodoo), they solve difficult problems in ways which are still not quite understood (building of the pyramids; Polynesian travels), there existed a highly developed and internationally known astronomy in the old Stone Age, this astronomy was factually adequate *as well as* emotionally satisfying, *it solved both physical and social problems* (one cannot say the same about modern astronomy) and it was tested in very simple and ingenious ways (stone observatories in England and in the South Pacific; astronomical schools in Polynesia - for a more detailed treatment and references concerning all these assertions c.f. my *Einführung in die Naturphilosophie*). There was the domestication of animals, the invention of rotating agriculture, new types of plants were bred and kept pure by careful avoidance of cross fertilization, we have chemical inventions, and we have a most amazing art that can compare with the best achievements of the present. True, there were no collective excursions to the moon, but single individuals, disregarding great dangers to their soul and their sanity, rose from sphere to sphere to sphere until they finally faced God himself in all His splendor while others changed into animals and back into humans again. At all times man approached his surroundings w' h wide open senses and a fertile intelligence, at all times he made incredible discoveries, at all times we can learn from his ideas.

Modern science, on the other hand, is not at all as difficult and as perfect as scientific propaganda wants us to believe. A subject such as medicine, or physics, or biology appears difficult only because it is taught badly, because the standard instructions are full of redundant material, and because they start too late in life. During the war, when the American Army needed physicians within a very short time, it was suddenly possible to reduce medical instruction to half a year (the corresponding instruction manuals have disappeared long ago, however. Science may be simplified during the war. In peacetime the prestige of science demands greater complication.) And how often does it not happen that the proud and conceited judgment of an expert is put in its proper place by a layman! Numerous inventors built 'impossible' machines. Lawyers show again and again that an expert does not know what he is talking about. Scientists, especially physicians, frequently come to different results so that it is up to the relatives of the sick person (or the inhabitants of a certain area) to decide *by vote* about the procedure to be adopted. How often is science improved, and turned into new directions by non-scientific influences! it is up to us, it is up to the citizens of a free society to either accept the chauvinism of science without contradiction or to overcome it by the counterforce of public action. Public action was used against science by the Communists in China in the fifties, and it was again used, under very different circumstances, by some opponents of evolution in California in the seventies. Let us follow their example and let us free society from the strangling hold of an ideologically petrified science just as our ancestors freed *us* from the strangling hold of the One True Religion!

The way towards this aim is clear. A science that insists on possessing the only correct method and the only acceptable results is ideology and must be separated from the state, and especially from the process of education. One may teach it, but only to those who have decided to make this particular superstition their own. On the other hand, a science that has dropped such totalitarian pretensions is no longer independent and self-contained, and it can be taught in many different combinations (myth and modern cosmology might be one such combination). Of course, every business has the right to demand that its practitioners be prepared in a special way, and it may even demand acceptance of a certain ideology (I for one am against the thinning out of subjects so that they become more and more similar to each other; whoever does not like present-day Catholicism should leave it and become a Protestant, or an Atheist, instead of ruining it by such inane changes as mass in the vernacular). That is true of physics, just as it is true of religion, or of prostitution. But such special ideologies, such special skills have no room in the process *of general education* that prepares a citizen for his role in society. A mature citizen is not a man who has been *instructed* in a special ideology, such as Puritanism, or critical rationalism, and who now carries this ideology with him like a mental tumor, a mature citizen is a person who has learned how to make up his mind and who has then *decided* in favor of what he thinks suits him best. He is a person who has a certain mental toughness (he does not fall for the first ideological street singer he happens to meet) and who is therefore able *consciously to choose* the business that seems to be most attractive to him rather than being swallowed by it. To prepare himself for his choice he will study the major ideologies as *historical phenomena,* he will study science as a historical phenomenon and not as the one and only sensible way of approaching a problem. He will study it together with other fairy-tales such as the myths of 'primitive' societies so that he has the information needed for arriving at a free decision. An essential part of a general education of this kind is acquaintance with the most outstanding propagandists in all fields, so that the pupil can build up his resistance against all propaganda, including the propaganda called 'argument'. It is only *after* such a hardening procedure that he will be called upon to make up his mind on the issue rationalism-irrationalism, science-myth, science-religion, and so on. His decision in favor of science - assuming he chooses science - will then be much more 'rational' than any decision in favor of science is today. At any rate - science and the schools will be just as carefully separated as religion and the schools are separated today. Scientists will of course participate in governmental decisions, for everyone participates in such decisions. But they will not be given overriding authority. It is the *vote of everyone concerned* that decides fundamental issues such as the teaching methods used, or the truth of basic beliefs such as the theory of evolution, or the quantum theory, and not the authority of big-shots hiding behind a non-existing methodology. There is no need to fear that such a way of arranging society will lead to undesirable results. Science itself uses the method of ballot, discussion, vote, thou-h without a clear grasp of its mechanism, and in a heavily biased way. But the rationality of our beliefs will certainly be considerably increased.

Chapter 9

ON AGGRESSION AND THE VIRTUE OF SCIENTIFIC HUMILITY

Konrad Lorenz

I may claim that the contents of the preceding chapters are natural science: the recorded facts are verified, as far as it is possible to say of the results of a science as young as comparative ethology. Now, however, we leave the record of facts elicited by observations and experiments on the aggressive behavior of animals and turn to the question of whether they can teach us something applicable to man and useful in circumventing the dangers arising from his aggressive drives.

There are people who see in this question an insult to human dignity. All too willingly man sees himself as the centre of the universe, as something not belonging to the rest of nature but standing apart as a different and higher being. Many people cling to this error and remain deaf to the wisest command ever given by a sage, the famous 'Know thyself' inscribed in the temple of Delphi. What keeps people from listening to it? There are three obstacles, all of them motivated by strong emotions. The first is easily overcome by the man of insight; the second is at least honorable, in spite of its harmful effects; the third is understandable from the standpoint of cultural history and is therefore forgivable, but it is the most difficult to remove. All three are inseparably bound up and shot through with a most dangerous human quality, of which the proverb says that it goes before a fall: pride. I will now discuss these obstacles and try to show in what manner they are harmful, and then I will do my best to contribute towards their elimination.

The first obstacle is the most primitive. It hinders self-knowledge in inhibiting man's awareness of his own evolutionary origin. Its irrational quality and its stubborn tenacity are paradoxically derived from the great likeness which our nearest animal relations bear to us. If people did not know the chimpanzee they would be more easily convinced of their own origin. An inexorable law of perception prevents us from seeing in the ape, particularly in the chimpanzee, an animal like other animals, and makes us see in its face the human physiognomy. From this point of view, measured by human standards, the chimpanzee of course appears as something horrible, a diabolical caricature of ourselves. In looking at the gorilla or the orangutan, which are less closely related to us, our judgment is correspondingly less distorted. The heads of the old males may look to us like bizarre devils' masks, impressive and even aesthetically appealing. However, we cannot feel like this about the chimpanzee: he is irresistibly funny and at the same time as common, as vulgar, as no other animal but a debased human being can ever be. This subjective impression is not altogether wrong: there are reasons for supposing that the common ancestor of man and the chimpanzee stood not lower but considerably higher than the chimpanzee does today. Absurd though the contemptuous attitude of man to the chimpanzee may be in itself, its strong emotional content has nevertheless misled several scientists into building up entirely unfounded theories about the origin of man: his evolution from animals is not disputed, but his close relationship to the repulsive chimpanzee is either passed over in a few logical skips or circumvented by sophistic detours.

The chimpanzee, however, is irresistibly funny just because he is so similar to us. What is worse is that in the narrow confinement of zoological gardens, adult chimpanzees degenerate much in the same way as human beings would under comparable circumstances, and give an impression of real dissoluteness and depravity. Even the normal chimp observed in perfect health gives the impression not of an extremely highly evolved animal but rather of a desperate and debased human being.

The second obstacle to self-knowledge is our reluctance to accept the fact that our own behavior obeys the laws of natural causation. Bernhard Hassenstein has called this attitude the 'anti-causal value judgment'. The reluctance of many people to recognize the causal determination of all natural phenomena, human behavior included, undoubtedly comes from the justifiable wish to possess a free will and to feel that our actions are determined not by fortuitous causes but by higher aims.

A third great obstacle to human self-knowledge is - at least in our Western cultures - a heritage of idealistic philosophy. It stems from the dichotomy of the world into the external world of things, which to idealistic thought is devoid of values, and the inner world of human thought and reason to which alone values are attributed. This division appeals to man's spiritual pride. It supports him in his reluctance to accept the determination of his own behavior by natural laws. How deeply it has penetrated into accepted ways of thinking can be seen from the alteration in meaning of the words 'idealist' and 'realist', which originally signified philosophic attitudes but today imply moral value judgments. We must realize how common it has become in Western, particularly German, thought to consider that whatever can be explained by the laws of nature is automatically devoid of higher values. To anybody thinking in this way explanation means devaluation.

I must here guard against the possible reproach that I am preaching against the three obstacles to human self-knowledge because they contradict my own scientific and philosophic views: I am not arguing against the rejection of the doctrine of evolution only because I am a convinced Darwinian; my opposition to the belief that natural explanation depreciates whatever it explains is not motivated by the fact that I happen to be professionally engaged in causal analysis; nor do I object to certain consequences of idealistic thought because my own epistemological attitude is that of hypothetical realism. I have better reasons. Science is often accused of having brought terrible dangers upon man by giving him too much power over nature. This accusation would be justified only if scientists were guilty of having neglected man himself as a subject for research. The danger to modern man arises not so much from his power of mastering natural phenomena as from his powerlessness to control sensibly what is happening today in his own society. I maintain that this powerlessness is entirely the consequence of the lack of human insight into the causation of human behavior. What I intend to show is that the insight necessary to control our own social behavior is blocked by the three pride-inspired obstacles to self-knowledge.

These obstacles prevent the causal analysis of all those processes in the life of man which he regards as being of particular value, in other words those processes of which he is proud. It cannot be stressed enough: the fact that the functions of our digestive system are well known and that, as a result of this knowledge, medicine, particularly intestinal surgery, saves many thousands of human lives annually, is entirely due to the fortunate circumstance that the functions of these organs do not evoke particular awe or reverence. If, on the other hand, we are powerless against the pathological disintegration of our social structure, and if, armed with atomic weapons, we cannot control our aggressive behavior any more sensibly than any animal species, this deplorable situation is largely due to our arrogant refusal to regard our own behavior as equally subject to the laws of nature and accessible to causal analysis.

Science is not to blame for men's lack of self-knowledge. Giordano Bruno went to the stake because he told his fellow men that they and their planet were only a speck of dust in a cloud of countless other specks. When Charles Darwin discovered that men are descended from animals they would have been glad to kill him, and there was certainly no lack of attempts to silence him. When Sigmund Freud attempted to analyze the motives of human social behavior and to explain its causes from the subjective-psychological side, but with the method of approach of true natural science, he was accused of irreverence, blind materialism and even pornographic tendencies. Humanity defends its own self-esteem with all its might, and it is certainly time to preach humility and try seriously to break down all obstructions to self-knowledge.

I will begin by attacking the resistance to the doctrine of Charles Darwin, and it may be considered an encouraging sign for the gradual spread of scientific education that today I no longer have to combat those who rose up against the findings of Giordano Bruno. I think I know a simple method of reconciling people to the fact that they are part of nature and have themselves originated by natural evolution without any infringement of natural laws: one need only show them the beauty and greatness of the universe, and the awe-inspiring laws that govern it. Surely nobody who knows enough about the phylogenetic evolution of the world of organisms can feel any inner resistance to the knowledge that he himself owes his existence to this greatest of all natural phenomena. I will not discuss the probability or rather the certainty of evolution, a certainty which by far surpasses that of all our historical knowledge. Everything we know confirms the fact of evolution; it possesses, too, everything that makes a 'myth of creation' valuable; utter convincingness, entrancing beauty and awe-inspiring greatness.

Anyone who understands this cannot possibly be repelled by Darwin's recognition of the fact that we have a common origin with animals, nor by Freud's realization that we are still driven by the same instincts as our prehuman ancestors. On the contrary, this knowledge inspires a new feeling of respect for the functions of reason and moral responsibility which first came into the world with man and which, provided he does not blindly and arrogantly deny the existence of his animal inheritance, give him the power to control it.

A further reason why some people still resist the doctrine of evolution is the great respect we human beings have for our ancestors. To descend from, means, literally, to come down, and even in Roman law it was customary to put the ancestor uppermost in the pedigree and to draw the family tree branching downwards. The fact that a human being has only two parents but 256 great-great-great-great-great-great-grand-parents does not appear in such pedigrees even if they extend to many generations. We avoid mentioning this multitude because among so many ancestors we would not find enough of whom we could be proud. According to some authors, the term 'descent' may derive from the fact that in ancient times man was fond of tracing his origin to the gods. That the family tree of life grows not from above downwards but from below upwards escaped man's notice until Darwin's time; thus the word 'descent' stands for the opposite of what it means, unless we wish to take it literally that our forefathers, in their time, came down from the trees. This they actually did, though as we know today, a long time before they became human beings.

The terms 'development' and 'evolution' are nearly as inadequate as 'descent'. They too came into use at a time when we knew nothing of the creative processes of the origin of species and only knew about the origin of individuals from eggs or seeds. A chick literally develops from an egg and a sunflower from a seed: that is nothing originates from the germ that was not already formed inside it. The growth of the great family tree of life is quite different. Though the ancestral form is the indispensable prerequisite for the origin of its more highly developed descendants, their evolution can in no way be predicted from the characters of the ancestor. The fact that birds evolved from dinosaurs or man from apes is a historically unique achievement of evolution. By laws that govern every living being evolution has a general trend to the higher but in all its details is determined by so-called chance that is by innumerable collateral chains of causation which in principle can never be completely apprehended. It is by 'chance' in this sense that from primitive forebears in Australia eucalyptus trees and kangaroos originated, and in Europe and Asia oak trees and man. The newly evolved form of life is an achievement, and its characters cannot be predicted from those of its forebear; that is, in the large majority of cases, something higher than the latter. The naive value-judgment expressed in the words 'lower animals' is for every unbiased person an inevitable necessity of thinking and feeling.

The scientist who considers himself absolutely 'objective' and believes that he can free himself from the compulsion of the 'merely' subjective should try - only in imagination of course to kill in succession a lettuce, a fly, a frog, a guinea pig, a cat, a dog, and finally a chimpanzee. He will then be aware how increasingly difficult murder becomes as the victim's level of organization rises. The degree of inhibition against killing each one of these beings is a very precise measure for the considerably different values that we cannot help attributing to lower and higher forms of life. To any man who finds it equally easy to chop up a live dog and a live lettuce I would recommend suicide at his earliest convenience!

The principle that science should be indifferent to values must not lead to the belief that evolution, that most wonderful of all chains of naturally explicable processes, is not capable of creating new values. That the origin of a higher form of life from a simpler ancestor means an increase in values is a reality as undeniable as that of our own existence.

None of our western languages has an intransitive verb to do justice to the increase of values produced by very nearly every step in evolution. One cannot possibly call it development when something new and higher arises from an earlier stage which does not contain the constituent properties for the new and higher being. Fundamentally this applies to each bigger step in the genesis of the world of organisms, including the first step, the origin of life, and the most recent one - the origin of man from the *Anthropoid.*

In spite of all epoch-making and inspiring new discoveries in biochemistry and virology, the origin of life is still the most puzzling of all natural phenomena. The difference between the processes of life and those occurring in non-living matter can only be defined by what Bernhard Hassenstein has termed an 'injunctive' definition. This means that to define the concept of life it is necessary to enumerate a number of constituent characteristics, none of which, taken by itself, constitutes life, but which, taken all together, in their summation and interaction, do indeed represent the essence of life. For each of the processes of life, such as metabolism, growth, and propagation and so on, an analogy can be found in inorganic matter, but all together can only be found in the living protoplasm. We are thus justified in maintaining that life processes are chemical and physical processes, and as such there is no doubt that fundamentally there is a natural explanation for them. No miracle is required to explain their peculiarities, for these are adequately explained by the complicated nature of molecular and other structures It is wrong, however, to assert that life processes are essentially chemical and physical processes. This assertion, though often made, contains unnoticed a false value judgment. The very 'essence' of life processes is that combination of characteristics which constitutes their *"injunctive" definition*, and it is with regard to these very characteristics that life processes are emphatically not what we ordinarily mean when we speak of chemical or physical processes. By virtue of the molecular structure of the living matter in which they take place the processes of life fulfill a great number of very particular functions, such as self-regulation, self-preservation, acquisition and storage of information and, above all, reproduction of the structures essential for these functions. These, though in principle causally explicable, cannot take place in other structurally less complex matter. In the world of organisms the relation of every higher life form to the lower one from which it originated is fundamentally the same as the relation of the processes and structures of life to those of the non-living. It would be a gross misrepresentation to say that the bird's wing is 'nothing but' a reptilian forelimb, or still worse, to say that man is 'nothing but' an ape. Indeed he is one, but he is much more besides: he is *essentially* more.

A sentimental misanthropist coined the often cited aphorism 'The more I see of human beings, the more I like animals". I maintain the contrary: only the person who knows animals, including the highest and most nearly related to ourselves, and who has gained insight into evolution, will be able to apprehend the unique position of man. We are the highest achievement reached so far by the great constructors of evolution. We are their 'latest' but certainly not their last word. The scientist must not regard anything as absolute, even the laws of pure reason. He must remain aware of the great fact, discovered by Heraclitus, that nothing whatever really remains the same even for one moment, but that everything is perpetually changing. To regard man, the most ephemeral and rapidly evolving of all species, as the final and unsurpassable achievement of creation, especially at his present-day particularly dangerous and disagreeable stage of development, is certainly the most arrogant and dangerous of all untenable doctrines. If I thought of man as the final image of God, I should not know what to think of God. But when I consider that our ancestors, at a time fairly recent in relation to the earth's history, were perfectly ordinary apes, closely related to chimpanzees, I see a glimmer of hope. It does not require very great optimism to assume that from us human beings something better and higher may evolve. Far from seeing in man the irrevocable and unsurpassable image of God, I assert - more modestly and, I believe, in greater awe of the Creation and its infinite possibilities- that the long-sought missing link between animals and the really human being is ourselves!

The first great obstacle to human self-knowledge, the reluctance to believe in our evolution from animals, is based, as I have tried to show, on ignorance or misunderstanding of the essence of organic creation. Fundamentally at least, it should be possible to remove this obstacle by teaching and learning. Similar means should help to remove the second obstacle, now to be discussed, the antipathy towards causal determination; but in this case the misunderstanding is far more difficult to clear up. Its root is the basically erroneous belief that a process which is causally determined cannot at the same time be goal-directed. Admittedly there are countless processes in the universe which are not goal-directed, and in these cases the question 'What for?' must remain unanswered, unless we are determined to find an answer at any price, in measureless overestimation of the importance of man, as for instance, if we explain the rising of the moon as a switching on of night illumination for our especial benefit. There is, however, no process to which the question of causes cannot be applied.

As already stated in Chapter III, the question 'What is it for?' only makes sense where the great constructors - or a living constructor constructed by them - have been at work. Only where parts of a systemic whole have become specialized, by division of labor, for different functions, each completing the other, does the question 'What is it for?' make any sense. This holds good for life processes, as also for those lifeless structures and functions which the living being makes use of for its own purposes, for instance, man-made machines. In these cases the question 'What is it for?' is not only relevant but absolutely necessary. We could not understand the cause of the cat's sharp claws if we had not first found out that their special function was catching mice.

At the beginning of Chapter VI, on the great parliament of instincts, we have already said that the answering of the question 'What is it for?' does not rule out the question of the cause. How little the two questions preclude each other can be shown by a simple analogy. I am driving through the countryside in my old car, to give a lecture in a distant town, and I ponder on the usefulness of my car, the goals or aims which are so well served by its construction, and it pleases me to think how all this contributes to achieve the purpose of my journey. Suddenly the motor coughs once or twice and peters out. At this stage I am painfully aware that the reason for my journey does not make my car go; I am learning the hard way that aims or goals are not causes. It will now be well for me to concentrate exclusively on the natural causes of the car's workings, and to find out at what stage the chain of their causation was so unpleasantly interrupted.

Medicine, 'queen of applied sciences', furnishes us even better examples of the erroneousness of the view that purposiveness and causality preclude each other. No 'life purpose', no 'whole-making factor' and no sense of imperative obligation can help the unfortunate patient with acute appendicitis, but even the youngest hospital surgeon can help him if he has rightly diagnosed the cause of the trouble. The appreciation of the fact that life processes are directed at aims or goals, and the realization of the other fact that they are, at the same time, determined by causality, not only do not preclude each other but they only make sense in combination. If man did not strive towards goals, his questions as to causes would have no sense; if he has no insight into cause and effect, he is powerless to guide effects towards determined goals, however rightly he may have understood the meaning of these goals.

This relation between the purposive and the causal aspects of life processes seems to me quite obvious, but evidently many people are under the illusion of their incompatibility. A classic example of how even a great mind can be a victim of this illusion is seen in the works of W. McDougall, the founder of 'purposive psychology'. In his book, *Outline of Psychology*, he rejects every causal physiological explanation of animal behavior, with one exception: he explains the malfunctioning of the light-compass-orientation, which causes insects to fly at night into flames, by so-called tropisms, or causally analyzed orientation mechanisms.

Probably the reason why people are so afraid of causal considerations is that they are terrified lest insight into the causes of earthly phenomena could expose man's free will as an illusion. In reality the fact that I have a will is as undeniable as the fact of my existence. Deeper insight into the physiological concatenation of causes of my own behavior cannot in the least alter the fact that I will but it can alter *what* I will.

Only on very superficial consideration does free will seem to imply that 'we can want what we will' in complete lawlessness, though this thought may appeal to those who flee as in claustrophobia from causality. We must remember how the theory of indeterminism of microphysical phenomena, the 'acausal' quantum physics, was avidly seized and on its foundations hypotheses built up to mediate between physical determinism and belief in free will, though the only freedom thereby left to the will was the lamentable liberty of the fortuitously cast die. Nobody can seriously believe that free will means that it is left entirely to the will of the individual, as to an irresponsible tyrant, to do or not do whatever he pleases. Our freest will underlies strict moral laws, and one of the reasons for our longing for freedom is to prevent our obeying other laws than these. It is significant that the anguished feeling of not being free is never evoked by the realization that our behavior is just as firmly bound to moral laws as physiological processes are to physical ones. We are all agreed that the greatest and most precious freedom of man is identical with the moral laws within him. Increasing knowledge of the natural causes of his own behavior can certainly increase a man's faculties and enable him to put his free will into action, but it can never diminish his will. If, in the impossible case of an utopian complete and ultimate success of causal analysis, man should ever achieve complete insight into the causality of earthly phenomena, including the workings of his own organism, he would not cease to have a will but it would be in perfect harmony with the incontrovertible lawfulness of the universe, the *Weltvernunft* of the Logos. This idea is foreign only to our present-day western thought; it was quite familiar to ancient Indian philosophy and to the mystics of the middle ages.

I now come to the third great obstacle to human self-knowledge, to the belief - deeply rooted in our western culture - that what can be explained in terms of natural science has no values. This belief springs from an exaggeration of Kant's values-philosophy, the consequence of the idealistic dichotomy of the world into the external world of things and the internal laws of human reason. As already intimated, fear of causality is one of the emotional reasons for the high values set on the unfathomable, but other unconscious factors are also involved. The behavior of the ruler, the father-figure, whose essential features include an element of arbitrariness and injustice, is unaccountable. God's decree is inscrutable. Whatever can be explained by natural causes can be controlled, and with its obscurity it loses most of its terror. Benjamin Franklin made of the thunderbolt, the instrument of Zeus's unaccountable whim, an electric spark against which the lightning conductors of our houses can protect us. The unfounded fear that nature might be desecrated by causal insight forms the second chief motive of people's fear of causality. Hence there arises a further obstacle to science, and this is all the stronger the greater a man's sense of the aesthetic beauty and awe-inspiring greatness of the universe and the more beautiful and venerable any particular natural phenomenon seems to him.

The obstacle to research arising from these unfortunate associations is the more dangerous since it never crosses the threshold of consciousness. If questioned, such people would profess in all sincerity to be supporters of scientific research, and within the limits of a circumscribed special field they may even be great scientists. But subconsciously they are firmly resolved not to carry their natural explanations beyond the limits of the awe-inspiring. Their error does not lie in the false assumption that some things are inexplicable: nobody knows so well as the scientist that there are limits to human understanding, but he is always aware that we do not know where these limits lie. Kant says, 'Our observation and analysis of its phenomena penetrate to the depth of nature. We do not know how far this will lead us in time.' The obstacle to scientific research produced by the attitude here discussed consists in setting a dogmatic border between the explorable and what is considered beyond exploration. Many excellent observers have so great a respect for life and its characteristics that they draw the line at its origin. They accept a special life force, *force vitale*, a direction-giving, whole-making factor which, they consider, neither requires nor permits an explanation. Others draw the line where they feel that human dignity demands a halt before any further attempts at natural explanation.

The attitude of the true scientist towards the real limits of human understanding was unforgettably impressed on me in early youth by the obviously unpremeditated words of a great biologist; Alfred Kuhn finished a lecture to the Austrian Academy of Sciences with Goethe's words, 'It is the greatest joy of the man of thought to have explored the explorable and then calmly to revere the inexplorable.' After the last word he hesitated, raised his hand in repudiation and cried, above the applause, 'No, not calmly, gentlemen; not *calmly*!' One could even define a true scientist by his ability to feel undiminished awe for the explorable that he has explored; from this arises his ability to want to explore the apparently inexplorable: he is not afraid of desecrating nature by causal insight. Never has natural explanation of one of its marvelous processes exposed nature as a charlatan who has lost the reputation of his sorcery; natural causal associations have always turned out to be grander and more awe-inspiring than even the most imaginative mythical interpretation. The true scientist does not need the inexplorable, the supernatural, to evoke his reverence: for him there is only one miracle, namely, that everything, even the finest flowerings of life, have come into being without miracles; for him the universe would lose some of its grandeur if he thought that any phenomenon even reason and moral sense in noble-minded human beings, could be accounted for only by an infringement of the omnipresent and omnipotent laws of one universe.

Nothing can better express the feelings of the scientist towards the great unity of the laws of nature than in Immanuel Kant's words: 'Two things fill the mind with ever new and increasing awe: the stars above me and the moral law within me.' Admiration and awe did not prevent the great philosopher from finding a natural explanation for the laws of the heavens, indeed an explanation based on their evolutionary origin. Would he, who did not yet know of the evolution of the world of organisms, be shocked that we consider the moral law within us not as something given, *a priori*, but as something which has arisen by natural evolution, just like the laws of the heavens?

Chapter 10

THE NATURE AND NECESSITY OF SCIENTIFIC REVOLUTIONS

Thomas Kuhn

These remarks permit us at last to consider the problems that provide this essay with its title. What are scientific revolutions, and what is their function in scientific development? Much of the answer to these questions has been anticipated in earlier sections. In particular, the preceding discussion has indicated that scientific revolutions are here taken to be those non-cumulative developmental episodes in which an older paradigm is replaced in whole or in part by an incompatible new one. There is more to be said, however, and an essential part of it can be introduced by asking one further question. Why should a change of paradigm be called a revolution? In the face of the vast and essential differences between political and scientific development, what parallelism can justify the metaphor that finds revolutions in both?

One aspect of the parallelism must already be apparent. Political revolutions are inaugurated by a growing sense, often restricted to a segment of the political community, that existing institutions have ceased adequately to meet the problems posed by an environment that they have in part created. In much the same way, scientific revolutions are inaugurated by a growing sense, again often restricted to a narrow subdivision of the scientific community that an existing paradigm has ceased to function adequately in the exploration of an aspect of nature to which that paradigm itself had previously led the way. In both political and scientific development the sense of malfunction that can lead to crisis is prerequisite to revolution. Furthermore, though it admittedly strains the metaphor that parallelism holds not only for the major paradigm changes, like those attributable to Copernicus and Lavoisier, but also for the far smaller ones associated with the assimilation of a new sort of phenomenon, like oxygen or X-rays. Scientific revolutions, as we noted at the end of Section V, need seem revolutionary only to those whose paradigms are affected by them. To outsiders they may, like the Balkan revolutions of the early twentieth century, seem normal parts of the developmental process. Astronomers, for example, could accept X-rays as a mere addition to knowledge, for their paradigms were unaffected by the existence of the new radiation. But for men like Kelvin, Crookes, and Roentgen, whose research dealt with radiation theory or with cathode ray tubes, the emergence of X-rays necessarily violated one paradigm as it created another. That is why these rays could be discovered only through something's first going wrong with normal research.

This genetic aspect of the parallel between political and scientific development should no longer be open to doubt. The parallel has, however, a second and more profound aspect upon which the significance of the first depends. Political revolutions aim to change political institutions in ways that those institutions themselves prohibit. Their success therefore necessitates the partial relinquishment of one set of institutions in favor of another, and in the interim, society is not fully governed by institutions at all. Initially it is crisis alone that attenuates the role of political institutions as we have already seen it attenuate the role of paradigms. In increasing numbers individuals become increasingly estranged from political life and behave more and more eccentrically within it. Then, as the crisis deepens, many of these individuals commit themselves to some concrete proposal for the reconstruction of society in a new institutional framework. At that point the society is divided into competing camps or parties, one seeking to defend the old institutional constellation, the others seeking to institute some new one. And, once that polarization has occurred, political recourse fails. Because they differ about the institutional matrix within which political change is to be achieved and evaluated, because they acknowledge no supra-institutional framework for the adjudication of revolutionary difference, the parties to a revolutionary conflict must finally resort to the techniques of mass persuasion, often including force. Though revolutions have had a vital role in the evolution of political institutions, that role depends upon their being partially extra political or extra institutional events. The remainder of this essay aims to demonstrate that the historical study of paradigm change reveals very similar characteristics in the evolution of the sciences. Like the choice between competing political institutions, that between competing paradigms proves to be a choice between incompatible modes of community life. Because it has that character, the choice is not and cannot be determined merely by the evaluative procedures characteristic of normal science, for these depend in part upon a particular paradigm, and that paradigm is at issue. When paradigms enter, as they must, into a debate about paradigm choice, their role is necessarily circular. Each group uses its own paradigm to argue in that paradigm's defense.

The resulting circularity does not, of course, make the arguments wrong or even ineffectual. The man who premises a paradigm when arguing in its defense can nonetheless provide a clear exhibit of what scientific practice will be like for those who adopt the new view of nature. That exhibit can be immensely persuasive, often compellingly so. Yet, whatever its force, the status of the circular argument is only that of persuasion. It cannot be made logically or even probabilistically compelling for those who refuse to step into the circle. The premises and values shared by the two parties to a debate over paradigms are not sufficiently extensive for that. As in political revolutions, so in paradigm choice – there is no standard higher than the assent of the relevant community. To discover how scientific revolutions are effected, we shall therefore have to examine not only the impact of nature and of logic, but also the techniques of persuasive argumentation effective within the quite special groups that constitute the community of scientists.

To discover why this issue of paradigm choice can never be unequivocally settled by logic and experiment alone, we must shortly examine the nature of the differences that separate the proponents of a traditional paradigm from their revolutionary successors. That examination is the principal object of this section and the next. We have, however, already noted numerous examples of such differences, and no one will doubt that history can supply many others. What is more likely to be doubted than their existence – and what must therefore be considered first – is that such examples provide essential information about the nature of science. Granting that paradigm rejection has been a historic fact, does it illuminate more than human credulity and confusion? Are there intrinsic reasons why the assimilation of either a new sort of phenomenon or a new scientific theory must demand the rejection of an older paradigm?

First notice that if there are such reasons, they do not derive from the logical structure of scientific knowledge. In principle, a new phenomenon might emerge without reflecting destructively upon any part of past scientific practice. Though discovering life on the moon would today be destructive of existing paradigms (these tell us things about the moon that seem incompatible with life's existence there), discovering life in some less well-known part of the galaxy would not. By the same token, a new theory does not have to conflict with any of its predecessors. It might deal exclusively with phenomena not previously known, as the quantum theory deals (but, significantly, not exclusively) with subatomic phenomena unknown before the twentieth century. Or again, the new theory might be simply a higher level theory than those known before, one that linked together a whole group of lower level theories without substantially changing any. Today, the theory of energy conservation provides just such links between dynamics, chemistry, electricity, optics, thermal theory, and so on. Still other compatible relationships between old and new theories can be conceived. Any and all of them might be exemplified by the historical process through which science has developed. If they were, scientific development would be genuinely cumulative. New sorts of phenomena would simply disclose order in an aspect of nature where none had been seen before. In the evolution of science new knowledge would replace ignorance rather than replace knowledge of another and incompatible sort.

Of course, science (or some other enterprise, perhaps less effective) might have developed in that fully cumulative manner. Many people have believed that it did so, and most still seem to suppose that accumulation is at least the ideal that historical development would display if only it had not so often been distorted by human idiosyncrasy. There are important reasons for that belief. In Section X we shall discover how closely the view of science-as-accumulation is entangled with a dominant epistemology that takes knowledge to be a construction placed directly upon raw sense data by the mind. And in Section XI we shall examine the strong support provided to the same historiographic schema by the techniques of effective science pedagogy. Nevertheless, despite the immense plausibility of that ideal image, there is increasing reason to wonder whether it can possibly be an image of science. After the pre-paradigm period the assimilation of all new theories and of almost all new sorts of phenomena has in fact demanded the destruction of a prior paradigm and a consequent conflict between competing schools of scientific thought. Cumulative acquisition of unanticipated novelties proves to be an almost non-existent exception to the rule of scientific development. The man who takes historic fact seriously must suspect that science does not tend toward the ideal that our image of its cumulativeness has suggested. Perhaps it is another sort of enterprise.

If, however, resistant facts can carry us that far, then a second look at the ground we have already covered may suggest that cumulative acquisition of novelty is not only rare in fact but improbable in principle. Normal research, which is cumulative, owes its success to the ability of scientists regularly to select problems that can be solved with conceptual and instrumental techniques close to those already in existence. (That is why an excessive concern with useful problems, regardless of their relation to existing knowledge and technique, can so easily inhibit scientific development.) The man who is striving to solve a problem defined by existing knowledge and technique is not, however, just looking around. He knows what he wants to achieve, and he designs his instruments and directs his thoughts accordingly. Unanticipated novelty, the new discovery, can emerge only to the extent that his anticipations about nature and his instruments prove wrong. Often the importance of the resulting discovery will itself be proportional to the extent and stubbornness of the anomaly that foreshadowed it. Obviously, then, there must be a conflict between the paradigm that discloses anomaly and the one that later renders the anomaly law-like. The examples of discovery through paradigm destruction examined in Section VI did not confront us with mere historical accident. There is no other effective way in which discoveries might be generated.

The same argument applies even more clearly to the invention of new theories. There are, in principle, only three types of phenomena about which a new theory might be developed. The first consists of phenomena already well explained by existing paradigms, and these seldom provide either motive or point of departure for theory construction. When they do, as with the three famous anticipations discussed at the end of Section VII, the theories that result are seldom accepted, because nature provides no ground for discrimination. A second class of phenomena consists of those whose nature is indicated by existing paradigms but whose details can be understood only through further theory articulation. These are the phenomena to which scientists direct their research much of the time, but that research aims at the articulation of existing paradigms rather than at the invention of new ones. Only when these attempts at articulation fail do scientists encounter the third type of phenomena, the recognized anomalies whose characteristic feature is their stubborn refusal to be assimilated to existing paradigms. This type alone gives rise to new theories. Paradigms provide all phenomena except anomalies with a theory-determined place in the scientist's field of vision.

But if new theories are called forth to resolve anomalies in the relation of an existing theory to nature, then the successful new theory must somewhere permit predictions that are different from those derived from its predecessor. That difference could not occur if the two were logically compatible. In the process of being assimilated, the second must displace the first. Even a theory like energy conservation, which today seems a logical superstructure that relates to nature only through independently established theories, did not develop historically without paradigm destruction. Instead, it emerged from a crisis in which an essential ingredient was the incompatibility between Newtonian dynamics and some recently formulated consequences of the caloric theory of heat. Only after the caloric theory had been rejected could energy conservation become part of science. And only after it had been part of science for some time could it come to seem a theory of a logically higher type, one not in conflict with its predecessors. It is hard to see how new theories could arise without these destructive changes in beliefs about nature. Though logical inclusiveness remains a permissible view of the relation between successive scientific theories, it is a historical implausibility.

Logical Positivism

A century ago it would, I think, have been possible to let the case for the necessity of revolutions rest at this point. But today, unfortunately, that cannot be done because the view of the subject developed above cannot be maintained if the most prevalent contemporary interpretation of the nature and function of scientific theory is accepted. That interpretation, closely associated with early logical positivism and not categorically rejected by its successors, would restrict the range and meaning of an accepted theory so that it could not possibly conflict with any later theory that made predictions about some of the same natural phenomena. The best-known and the strongest case for this restricted conception of a scientific theory emerges in discussions of the relation between contemporary Einsteinian dynamics and the older dynamical equations that descend from Newton's *Principia*. From the viewpoint of this essay these two theories are fundamentally incompatible in the sense illustrated by the relation of Copernican to Ptolemaic astronomy: Einstein's theory can be accepted only with the recognition that Newton's was wrong. Today this remains a minority view. We must therefore examine the most prevalent objections to it.

The gist of these objections can be developed as follows. Relativistic dynamics cannot have shown Newtonian dynamics to be wrong, for Newtonian dynamics is still used with great success by most engineers and, in selected applications, by many physicists. Furthermore, the propriety of this use of the older theory can be proved from the very theory that has, in other applications, replaced it. Einstein's theory can be used to show that predictions from Newton's equations will be as good as our measuring instruments in all applications that satisfy a small number of restrictive conditions. For example, if Newtonian theory is to provide a good approximate solution, the relative velocities of the bodies considered must be small compared with the velocity of light. Subject to this condition and a few others, Newtonian theory seems to be derivable from Einsteinian, of which it is therefore a special case.

But, the objection continues, no theory can possibly conflict with one of its special cases. If Einsteinian science seems to make Newtonian dynamics wrong, that is only because some Newtonians were so incautious as to claim that Newtonian theory yielded entirely precise results or that it was valid at very high relative velocities. Since they could not have had any evidence for such claims, they betrayed the standards of science when they made them. In so far as Newtonian theory was ever a truly scientific theory supported by valid evidence, it still is. Only extravagant claims for the theory – claims that were never properly parts of science can have been shown by Einstein to be wrong. Purged of these merely human extravagances, Newtonian theory has never been challenged and cannot be.

Some variant of this argument is quite sufficient to make any theory ever used by a significant group of competent scientists immune to attack. The much-maligned phlogiston theory, for example, gave order to a large number of physical and chemical phenomena. It explained why bodies burned – they were rich in phlogiston – and why metals had so many more properties in common than did their ores. The metals were all compounded from different elementary earths combined with phlogiston, and the latter, common to all metals, produced common properties. In addition, the phlogiston theory accounted for a number of reactions in which acids were formed by the combustion of substances like carbon and sulphur. Also, it explained the decrease of volume when combustion occurs in a confined volume of air the phlogiston released by combustion "spoils" the elasticity of the air that absorbed it, just as fire "spoils" the elasticity of a steel spring. If these were the only phenomena that the phlogiston theorists had claimed for their theory, that theory could never have been challenged. A similar argument will suffice for any theory that has ever been successfully applied to any range of phenomena at all.

But to save theories in this way, their range of application must be restricted to those phenomena and to that precision of observation with which the experimental evidence in hand already deals. Carried just a step further (and the step can scarcely be avoided once the first is taken), such a limitation prohibits the scientist from claiming to speak "scientifically" about any phenomenon not already observed. Even in its present form the restriction forbids the scientist to rely upon a theory in his own research whenever that research enters an area or seeks a degree of precision for which past practice with the theory offers no precedent. These prohibitions are logically unexceptionable. But the result of accepting them would be the end of the research through which science may develop further.

By now that point too is virtually a tautology. Without commitment to a paradigm there could be no normal science. Furthermore, that commitment must extend to areas and to degrees of precision for which there is no full precedent. If it did not, the paradigm could provide no puzzles that had not already been solved. Besides, it is not only normal science that depends upon commitment to a paradigm. If existing theory binds the scientist only with respect to existing applications, then there can be no surprises, anomalies, or crises. But these are just the signposts that point the way to extraordinary science. If positivistic restrictions on the range of a theory's legitimate applicability are taken literally, the mechanism that tells the scientific community what problems may lead to fundamental change must cease to function. And when that occurs, the community will inevitably return to something much like its pre-paradigm state a condition in which all members practice science but in which their gross product scarcely resembles science at all. Is it really any wonder that the price of significant scientific advance is a commitment that runs the risk of being wrong?

More important, there is a revealing logical lacuna in the positivist's argument, one that will reintroduce us immediately to the nature of revolutionary change. Can Newtonian dynamics really be derived from relativistic dynamics? What would such a derivation look like? Imagine a set of statements, E_1, E_2 ... E_n which together embody the laws of relativity theory. These statements contain variables and parameters representing spatial position, time, rest mass, etc. From them, together with the apparatus of logic and mathematics, is deducible a whole set of further statements including some that can be checked by observation. To prove the adequacy of Newtonian dynamics as a special case, we must add to the E_i's additional statements, like $(v/c)^2 << 1$, restricting the range of the parameters and variables. This enlarged set of statements is then manipulated to yield a new set, N_1, N_2... N_m, which is identical in form with Newton's laws of motion, the law of gravity, and so on. Apparently Newtonian dynamics has been derived from Einsteinian, subject to a few limiting conditions.

Yet the derivation is spurious, at least to this point. Though the N_i's are a special case of the laws of relativistic mechanics, they are not Newton's Laws.

Or at least they are not unless those laws are reinterpreted in a way that would have been impossible until after Einstein's work. The variables and parameters that in the Einsteinian E_i's represented spatial position, time, mass, etc., still occur in the N_i's; and they there still represent Einsteinian space, time, and mass. But the physical referents of these Einsteinian concepts are by no means identical with those of the Newtonian concepts that bear the same name. (Newtonian mass is conserved; Einsteinian is convertible with energy. Only at low relative velocities may the two be measured in the same way, and even then they must not be conceived to be the same.) Unless we change the definitions of the variables in the N_i's, the statements we have derived are not Newtonian. If we do change them, we cannot properly be said to have derived Newton's Laws, at least not in any sense of "derive" now generally recognized. Our argument has, of course, explained why Newton's Laws ever seemed to work. In doing so it has justified, say, an automobile driver in acting as though he lived in a Newtonian universe. An argument of the same type is used to justify teaching earth-centered astronomy to surveyors. But the argument has still not done what it purported to do. It has not, that is, shown Newton's Laws to be a limiting case of Einstein's. For in the passage to the limit it is not only the forms of the laws that have changed. Simultaneously we have had to alter the fundamental structural elements of which the universe to which they apply is composed.

This need to change the meaning of established and familiar concepts is central to the revolutionary impact of Einstein's theory. Though subtler than the changes from geocentrism to heliocentrism, from phlogiston to oxygen, or from corpuscles to waves, the resulting conceptual transformation is no less decisively destructive of a previously established paradigm. We may even come to see it as a prototype for revolutionary reorientations in the sciences. Just because it did not involve the introduction of additional objects or concepts, the transition from Newtonian to Einsteinian mechanics illustrates with particular clarity the scientific revolution as a displacement of the conceptual network through which scientists view the world.

These remarks should suffice to show what might, in another philosophical climate, have been taken for granted. At least for scientists, most of the apparent differences between a discarded scientific theory and its successor are real. Though an out-of-date theory can always be viewed as a special case of its up-to-date successor, it must be transformed for the purpose. And the transformation is one that can be undertaken only with the advantages of hindsight, the explicit guidance of the more recent theory. Furthermore, even if that transformation were a legitimate device to employ in interpreting the older theory, the result of its application would be a theory so restricted that it could only restate what was already known. Because of its economy, that restatement would have utility, but it could not suffice for the guidance of research.

Let us, therefore, now take it for granted that the differences between successive paradigms are both necessary and irreconcilable. Can we then say more explicitly what sorts of differences these are? The most apparent type has already been illustrated repeatedly. Successive paradigms tell us different things about the population of the universe and about that population's behavior. They differ, that is, about such questions as the existence of subatomic particles, the materiality of light, and the conservation of heat or of energy. These are the substantive differences between successive paradigms, and they require no further illustration. But paradigms differ in more than substance, for they are directed not only to nature but also back upon the science that produced them. They are the source of the methods, problem-field, and standards of solution accepted by any mature scientific community at any given time. As a result, the reception of a new paradigm often necessitates a redefinition of the corresponding science. Some old problems may be relegated to another science or declared entirely "unscientific." Others that were previously non-existent or trivial may, with a new paradigm, become the very archetypes of significant scientific achievement. And as the problems change, so, often, does the standard that distinguishes a real scientific solution from a mere metaphysical speculation, word game, or mathematical play. The normal-scientific tradition that emerges from a scientific revolution is not only incompatible but often actually incommensurable with that which has gone before.

The impact of Newton's work upon the normal seventeenth century tradition of scientific practice provides a striking example of these subtler effects of paradigm shift. Before Newton was born the "new science" of the century had at last succeeded in rejecting Aristotelian and scholastic explanations expressed in terms of the essences of material bodies. To say that a stone fell because its "nature" drove it toward the center of the universe had been made to look a mere tautological word-play, something it had not previously been. Henceforth the entire flux of sensory appearances, including color, taste, and even weight, was to be explained in terms of the size, shape, position, and motion of the elementary corpuscles of base matter. The attribution of other qualities to the elementary atoms was a resort to the occult and therefore out of bounds for science. Molière caught the new spirit precisely when he ridiculed the doctor who explained opium's efficacy as a soporific by attributing to it a dormitive potency. During the last half of the seventeenth century many scientists preferred to say that the round shape of the opium particles enabled them to sooth the nerves about which they moved.

In an earlier period explanations in terms of occult qualities had been an integral part of productive scientific work. Nevertheless, the seventeenth century's new commitment to mechanico-corpuscular explanation proved immensely fruitful for a number of sciences, ridding them of problems that had defied generally accepted solution and suggesting others to replace them. In dynamics, for example, Newton's three laws of motion are less a product of novel experiments than of the attempt to reinterpret well-known observations in terms of the motions and interactions of primary neutral corpuscles. Consider just one concrete illustration. Since neutral corpuscles could act on each other only by contact, the mechanico-corpuscular view of nature directed scientific attention to a brand-new subject of study, the alteration of particulate motions by collisions. Descartes announced the problem and provided its first putative solution. Huygens, Wren, and Wallis carried it still further, partly by experimenting with colliding pendulum bobs, but mostly by applying previously well-known characteristics of motion to the new problem. And Newton embedded their results in his laws of motion. The equal "action" and "reaction" of the third law are the changes in quantity of motion experienced by the two parties to a collision. The same change of motion supplies the definition of dynamical force implicit in the second law. In this case, as in many others during the seventeenth century, the corpuscular paradigm bred both a new problem and a large part of that problem's solution.

Yet, though much of Newton's work was directed to problems and embodied standards derived from the mechanico-corpuscular world view, the effect of the paradigm that resulted from his work was a further and partially destructive change in the problems and standards legitimate for science. Gravity, interpreted as an innate attraction between every pair of particles of matter, was an occult quality in the same sense as the scholastics' "tendency to fall" had been. Therefore, while the standards of corpuscularism remained in effect, the search for a mechanical explanation of gravity was one of the most challenging problems for those who accepted the *Principia* as paradigm. Newton devoted much attention to it and so did many of his eighteenth-century successors. The only apparent option was to reject Newton's theory for its failure to explain gravity, and that alternative, too, was widely adopted. Yet neither of these views ultimately triumphed. Unable either to practice science without the *Principia* or to make that work conform to the corpuscular standards of the seventeenth century, scientists gradually accepted the view that gravity was indeed innate. By the mid-eighteenth century that interpretation had been almost universally accepted, and the result was a genuine reversion (which is not the same as a retrogression) to a scholastic standard. Innate attractions and repulsions joined size, shape, position, and motion as physically irreducible primary properties of matter.

The resulting change in the standards and problem-field of physical science was once again consequential. By the 1740's, for example, electricians could speak of the attractive "virtue" of the electric fluid without thereby inviting the ridicule that had greeted Molière's doctor a century before. As they did so, electrical phenomena increasingly displayed an order different from the one they had shown when viewed as the effects of a mechanical effluvium that could act only by contact. In particular, when electrical action-at-a-distance became a subject for study in its own right, the phenomenon we now call charging by induction could be recognized as one of its effects. Previously, when seen at all, it had been attributed to the direct action of electrical "atmospheres" or to the leakages inevitable in any electrical laboratory. The new view of inductive effects was, in turn, the key to Franklin's analysis of the Leyden jar and thus to the emergence of a new and Newtonian paradigm for electricity. Nor were dynamics and electricity the only scientific fields affected by the legitimization of the search for forces innate to matter. The large body of eighteenth-century literature on chemical affinities and replacement series also derives from this supramechanical aspect of Newtonianism. Chemists who believed in these differential attractions between the various chemical species set up previously unimagined experiments and searched for new sorts of reactions. Without the data and the chemical concepts developed in that process, the later work of Lavoisier and, more particularly, of Dalton would be incomprehensible. Changes in the standards governing permissible problems, concepts, and explanations can transform a science. In the next section I shall even suggest a sense in which they transform the world.

Other examples of these non-substantive differences between successive paradigms can be retrieved from the history of any science in almost any period of its development. For the moment let us be content with just two other and far briefer illustrations. Before the chemical revolution, one of the acknowledged tasks of chemistry was to account for the qualities of chemical substances and for the changes these qualities underwent during chemical reactions. With the aid of a small number of elementary "principles" – of which phlogiston was one – the chemist was to explain why some substances are acidic, others metalline, combustible, and so forth. Some success in this direction had been achieved. We have already noted that phlogiston explained why the metals were so much alike, and we could have developed a similar argument for the acids. Lavoisier's reform, however, ultimately did away with chemical "principles," and thus ended by depriving chemistry of some actual and much potential explanatory power. To compensate for this loss, a change in standards was required. During much of the nineteenth century failure to explain the qualities of compounds was no indictment of a chemical theory.

Or again, Clerk Maxwell shared with other nineteenth-century proponents of the wave theory of light the conviction that light waves must be propagated through a material ether. Designing a mechanical medium to support such waves was a standard problem for many of his ablest contemporaries. His own theory, however, the electromagnetic theory of light, gave no account at all of a medium able to support light waves, and it clearly made such an account harder to provide than it had seemed before. Initially, Maxwell's theory was widely rejected for those reasons. But, like Newton's theory, Maxwell's proved difficult to dispense with, and as it achieved the status of a paradigm the community's attitude toward it changed. In the early decades of the twentieth century Maxwell's insistence upon the existence of a mechanical ether looked more and more like lip service, which it emphatically had not been, and the attempts to design such an ethereal medium were abandoned. Scientists no longer thought it unscientific to speak of an electrical "displacement" without specifying what was being displaced. The result, again, was a new set of problems and standards, one which, in the event, had much to do with the emergence of relativity theory.

These characteristic shifts in the scientific community's conception of its legitimate problems and standards would have less significance to this essay's thesis if one could suppose that they always occurred from some methodologically lower to some higher type. In that case their effects, too, would seem cumulative. No wonder that some historians have argued that the history of science records a continuing increase in the maturity and refinement of man's conception of the nature of science. Yet the case for cumulative development of science's problems and standards is even harder to make than the case for cumulation of theories. The attempt to explain gravity, though fruitfully abandoned by most eighteenth-century scientists, was not directed to an intrinsically illegitimate problem; the objections to innate forces were neither inherently unscientific nor metaphysical in some pejorative sense. There are no external standards to permit a judgment of that sort. What occurred was neither a decline nor a raising of standards, but simply a change demanded by the adoption of a new paradigm. Furthermore, that change has since been reversed and could be again. In the twentieth century Einstein succeeded in explaining gravitational attractions and that explanation has returned science to a set of canons and problems that are, in this particular respect, more like those of Newton's predecessors than of his successors. Or again, the development of quantum mechanics has reversed the methodological prohibition that originated in the chemical revolution. Chemists now attempt, and with great success, to explain the color, state of aggregation, and other qualities of the substances used and produced in their laboratories. A similar reversal may even be underway in electromagnetic theory. Space, in contemporary physics, is not the inert and homogenous substratum employed in both Newton's and Maxwell's theories; some of its new properties are not unlike those once attributed to the ether; we may someday come to know what an electric displacement is.

By shifting emphasis from the cognitive to the normative functions of paradigms, the preceding examples enlarge our understanding of the ways in which paradigms give form to the scientific life. Previously, we had principally examined the paradigm's role as a vehicle for scientific theory. In that role it functions by telling the scientist about the entities that nature does and does not contain and about the ways in which those entities behave. That information provides a map whose details are elucidated by mature scientific research. And since nature is too complex and varied to be explored at random, that map is as essential as observation and experiment to science's continuing development. Through the theories they embody, paradigms prove to be constitutive of the research activity. They are also however, constitutive of science in other respects, and that is now the point. In particular, our most recent examples show that paradigms provide scientists not only with a map but also with some of the directions essential for map-making. In learning a paradigm the scientist acquires theory, methods, and standards together, usually in an inextricable mixture. Therefore, when paradigms change, there are usually significant shifts in the criteria determining the legitimacy both of problems and of proposed solutions.

That observation returns us to the point from which this section began, for it provides our first explicit indication of why the choice between competing paradigms regularly raises questions that cannot be resolved by the criteria of normal science. To the extent, as significant as it is incomplete, that two scientific schools disagree about what is a problem and what a solution, they will inevitably talk through each other when debating the relative merits of their respective paradigms. In the partially circular arguments that regularly result, each paradigm will be shown to satisfy more or less the criteria that it dictates for itself and to fall short of a few of those dictated by its opponent. There are other reasons; too, for the incompleteness of logical contact that consistently characterizes paradigm debates. For example, since no paradigm ever solves all the problems it defines and since no two paradigms leave all the same problems unsolved, paradigm debates always involve the question: Which problems is it more significant to have solved? Like the issue of competing standards, that question of values can be answered only in terms of criteria that lie outside of normal science altogether, and it is that recourse to external criteria that most obviously makes paradigm debates revolutionary. Something even more fundamental than standards and values is, however, also at stake. I have so far argued only that paradigms are constitutive of science. Now I wish to display a sense in which they are constitutive of nature as well.

Postscript: Revolutions and Relativism

One consequence of the position just outlined has particularly bothered a number of my critics. They find my viewpoint relativistic, particularly as it is developed in the last section of this book. My remarks about translation highlight the reasons for the charge. The proponents of different theories are like the members of different language-culture communities. Recognizing the parallelism suggests that in some sense both groups may be right. Applied to culture and its development that position is relativistic.

But applied to science it may not be, and it is in any case far from mere relativism in a respect that its critics have failed to see. Taken as a group or in groups, practitioners of the developed sciences are, I have argued, fundamentally puzzle-solvers. Though the values that they deploy at times of theory-choice derive from other aspects of their work as well, the demonstrated ability to set up and to solve puzzles presented by nature is, in case of value conflict, the dominant criterion for most members of a scientific group. Like any other value, puzzle-solving ability proves equivocal in application. Two men who share it may nevertheless differ in the judgments they draw from its use. But the behavior of a community which makes it pre-eminent will be very different from that of one which does not. In the sciences, I believe, the high value accorded to puzzle-solving ability has the following consequences.

Imagine an evolutionary tree representing the development of the modern scientific specialties from their common origins in, say, primitive natural philosophy and the crafts. A line drawn up that tree, never doubling back, from the trunk to the tip of some branch would trace a succession of theories related by descent. Considering any two such theories, chosen from points not too near their origin, it should be easy to design a list of criteria that would enable an uncommitted observer to distinguish the earlier from the more recent theory time after time. Among the most useful would be: accuracy of prediction, particularly of quantitative prediction; the balance between esoteric and everyday subject matter; and the number of different problems solved. Less useful for this purpose, though also important determinants of scientific life would be such values as simplicity, scope, and compatibility with other specialties. Those lists are not yet the ones required, but I have no doubt that they can be completed. If they can, then scientific development is, like biological, a unidirectional and irreversible process. Later scientific theories are better than earlier ones for solving puzzles in the often quite different environments to which they are applied. That is not a relativist's position, and it displays the sense in which I am a convinced believer in scientific progress.

Compared with the notion of progress most prevalent among both philosophers of science and laymen, however, this position lacks an essential element. A scientific theory is usually felt to be better than its predecessors not only in the sense that it is a better instrument for discovering and solving puzzles but also because it is somehow a better representation of what nature is really like. One often hears that successive theories grow ever closer to, or approximate more and more closely to, the truth. Apparently generalizations like that refer not to the puzzle-solutions and the concrete predictions derived from a theory but rather to its ontology, to the match, that is, between the entities with which the theory populates nature and what is "really there."

Perhaps there is some other way of salvaging the notion of 'truth' for application to whole theories, but this one will not do. There is, I think, no theory-independent way to reconstruct phrases like 'really there'; the notion of a match between the ontology of a theory and its "real" counterpart in nature now seems to me illusive in principle. Besides, as a historian, I am impressed with the implausibility of the view. I do not doubt, for example, that Newton's mechanics improves on Aristotle's and that Einstein's improves on Newton's as instruments for puzzle-solving. But I can see in their succession no coherent direction of ontological development. On the contrary, in some important respects, though by no means in all, Einstein's general theory of relativity is closer to Aristotle's than either of them is to Newton's. Though the temptation to describe that position as relativistic is understandable, the description seems to me wrong. Conversely, if the position be relativism, I cannot see that the relativist loses anything needed to account for the nature and development of the sciences ...

INDEX

C

D

E

F

G

H

I

J

K

L

M

N

O

P

Q

R

S

T

U

V

W

Y

Z